STUDIES IN APPLIED MECHANICS 11

Mechanics of Material Interfaces

STUDIES IN APPLIED MECHANICS

1. Mechanics and Strength of Materials (Skalmierski)
2. Nonlinear Differential Equations (Fučík and Kufner)
3. Mathematical Theory of Elastic and Elastico-Plastic Bodies
   An Introduction (Necas and Hlaváček)
4. Variational, Incremental and Energy Methods in Solid Mechanics
   and Shell Theory (Mason)
5. Mechanics of Structured Media, Parts A and B (Selvadurai, Editor)
6. Mechanics of Material Behavior (Dvorak and Shield, Editors)
7. Mechanics of Granular Materials: New Models and Constitutive Relations
   (Jenkins and Satake, Editors)
8. Probabilistic Approach to Mechanisms (Sandler)
9. Methods of Functional Analysis for Application in Solid Mechanics (Mason)
10. Boundary Integral Equation Methods in Eigenvalue Problems of Elastodynamics
    and Thin Plates (Kitahara)
11. Mechanics of Material Interfaces (Selvadurai and Voyiadjis, Editors)
12. Local Effects in the Analysis of Structures (Ladevèze, Editor)

STUDIES IN APPLIED MECHANICS 11

Mechanics of Material Interfaces

Proceedings of the Technical Sessions on Mechanics of Material Interfaces held at the ASCE/ASME Mechanics Conference, Albuquerque, New Mexico, June 23–26, 1985

Edited by

A.P.S. Selvadurai

Department of Civil Engineering, Carleton University, Ottawa, Ontario, Canada

and

G.Z. Voyiadjis

Department of Civil Engineering, Louisiana State University, Baton Rouge, Louisiana, U.S.A.

Technical Sessions Sponsored by the Elasticity Committee of the Engineering Mechanics Division of the American Society of Civil Engineers

ELSEVIER
Amsterdam – Oxford – New York – Tokyo 1986

ELSEVIER SCIENCE PUBLISHERS B.V.
Sara Burgerhartstraat 25
P.O. Box 211, 1000 AE Amsterdam, The Netherlands

Distributors for the United States and Canada:

ELSEVIER SCIENCE PUBLISHING COMPANY INC.
52, Vanderbilt Avenue
New York, NY 10017, U.S.A.

Library of Congress Cataloging-in-Publication Data

Joint ASCE/ASME Mechanics Conference (1985 :
Albuquerque, N.M.)
Mechanics of material interfaces.

(Studies in applied mechanics ; 11)
"Technical sessions sponsored by the Elasticity Committee of the Engineering Mechanics Division of the American Society of Civil Engineers."
Bibliography: p.
Includes index.
1. Surfaces (Technology)--Congresses. 2. Mechanics--Congresses. I. Selvadurai, A. P. S. II. Voyiadjis, G. Z. III. American Society of Civil Engineers. Engineering Mechanics Division. Elasticity Committee. IV. Title. V. Series.
TA407.J68 1985 620.1'05 86-4467
ISBN 0-444-41758-3

ISBN 0-444-42625-6 (Vol. 11)
ISBN 0-444-41758-3 (Series)

Printed in The Netherlands.

PREFACE

The category of problems that examines the mechanical behaviour of contact regions constitutes an important branch of applied mechanics with extensive engineering applications. The results of such research can be applied to the study of mechanics of composite materials, tribology, soil-foundation interaction, mechanics of rock interfaces, modelling of damage phenomena and micro-mechanics. In classical studies, the modelling of interface responses has focussed on purely idealized forms of interface phenomena which range from frictionless contact to bonded contact, with Coulomb friction or finite friction occupying an intermediate position. Current research has attempted to improve such modelling by endowing the interface with its own characteristic constitutive responses. These recent researches indicate the significant manner in which non linear, frictional, dilatant, hardening and softening interface constitutive responses can influence the global and local interface responses of engineering interest.

The Technical sessions on *Mechanics of Material Interfaces* held at the ASCE/ASME Mechanics Conference, Albuquerque, New Mexico, from the 23rd to 26th June 1985, were sponsored by the Elasticity Committee of the Engineering Mechanics Division of the American Society of Civil Engineers. The objectives of these technical sessions were to bring together recent advances in the theoretical formulation, analysis and the application of material interface modelling to problems of engineering interest. The current volume contains those papers which were presented at the Conference and certain invited contributions from leading researchers. These papers focus on a variety of themes including fundamental phenomena in the modelling of frictional contact, modelling of interfaces in geomechanics, contact and inclusion problems and the study of the influences of interface phenomena in fracture mechanics.

The Editors wish to thank the authors for their willingness to both contribute to the volume and for their participation at the Technical Sessions at the ASCE/ASME meeting. The Editors gratefully acknowledge the support and encouragement of Professor L.M. Brock, (University of Kentucky) the former Chairman of the Elasticity Committee of the Engineering Mechanics Division of the ASCE, which led to the organization of the Technical Sessions and to the compilation of the Proceedings in this volume form. Thanks are also due to Mrs. S.J. Selvadurai for the editorial assistance and for the preparation of the index of keywords.

A.P.S. Selvadurai, Ottawa, Ontario, Canada
G.Z. Voyiadjis, Baton Rouge, LA, U.S.A.

LIST OF CONTRIBUTORS

A. Azarkhin, ALCOA Technical Centre, PA, U.S.A.

J.R. Barber, University of Michigan, MI, U.S.A.

T. Belytschko, Northwestern University, IL, U.S.A.

L.M. Brock, University of Kentucky, KY, U.S.A.

E.P. Chen, Sandia National Laboratories, NM, U.S.A.

M. Comninou, University of Michigan, MI, U.S.A.

S.K. Datta, University of Colorado, CO, U.S.A.

C.S. Desai, University of Arizona, AZ, U.S.A.

M.O. Faruque, University of Rhode Island, RI, U.S.A.

A. Klarbring, Linköping Institute of Technology, Linköping, Sweden

H.M. Ledbetter, National Bureau of Standards, CO, U.S.A.

N. Li, University of Michigan, MI, U.S.A.

J.A.C. Martins, The University of Texas at Austin, TX, U.S.A.

F.A. Mirza, McMaster University, Ont, Canada

B.K. Nagaraj, University of Arizona, AZ, U.S.A.

J.T. Oden, The University of Texas at Austin, TX, U.S.A.

P.D. Panagiotopoulos, Aristotle University Thessaloniki, Greece

M. Plesha, University of Wisconsin, WI, U.S.A.

A.A. Poe, Louisiana State University, LA, U.S.A.

R.K.N.D. Rajapakse, Carleton University, Ont, Canada

A.P.S. Selvadurai, Carleton University, Ont, Canada

D.F.E. Stolle, McMaster University, Ont, Canada

G.Z. Voyiadjis, Louisiana State University, LA, U.S.A.

TABLE OF CONTENTS

FUNDAMENTAL PHENOMENA AT INTERFACES WITH FRICTIONAL CONTACT

INTERFACE MODELS, VARIATIONAL PRINCIPLES AND NUMERICAL SOLUTIONS FOR DYNAMIC FRICTION PROBLEMS

J.A.C. MARTINS and J.T. ODEN

Texas Institute for Computational Mechanics, Aerospace Engineering and Engineering Mechanics Department, The University of Texas at Austin

ABSTRACT

The present paper summarizes several results of our current research on dynamic friction problems involving dry metallic bodies. We develop a model of interface response which incorporates a constitutive equation for the normal deformability of the interface and the Coulomb's law of friction. We develop continuum models, variational principles and finite elements approximations for elastodynamics and steady sliding frictional problems. Existence and uniqueness for the latter hold under appropriate assumptions on the data. The dynamic stability of steady frictional sliding is numerically studied and, assuming no distinction between static and kinetic coefficients of friction, we demonstrate numerically the possibility of the occurrence of apparent distinctions between those coefficients and the occurrence of stick-slip relaxation oscillations.

INTRODUCTION. AN INTERFACE MODEL

Stick-slip motion vs. steady sliding

Let us consider a body (slider) in frictional contact with a fixed surface, as depicted in Fig. 1a. The slider is intended to move along the surface with some (small) tangential velocity $\dot{U}_T$, this velocity being impressed upon it by some driving mechanism which itself possesses specific elastic and damping properties. As many examples in our daily life teach us, what frequently happens is that the desired steady sliding at the prescribed velocity $\dot{U}_T$ is not achieved. Instead, the body advances in an intermittent fashion with successive alternate periods of rest and sudden sliding (see Fig. 2a). These relaxation type oscillations are usually known as stick-slip motion (after Bowden and Leben (ref.2)).

Since the work of Blok (ref.3) in 1940, it is known that the essential condition for the occurrence of stick-slip oscillations in sliding motion is a decrease of the friction force with the increase, from zero, of the sliding velocity. Since then, most of the experimental studies of the stick-slip oscillations have concentrated on the determination of the functional dependence of the coefficient of friction with the sliding velocity. The overwhelming majority of such experimental studies involve experimental methods in which the tangential motion and the tangential forces on a slider, such as that in Fig. 1, are carefully monitored while the net normal force is assumed to be constant. Then, all the variations in the observed friction force are assumed to be a consequence of corresponding variations of the coefficient of friction with sliding velocity.

Unfortunately, the results of literally hundreds of experimental tests have little, if any, agreement. While it may be accepted that different pairs of metals in contact and different lubrication conditions produce characteristics (coefficient of friction vs. sliding velocity plots) with clearly distinct shapes, it is known that even for the same combination of metals and lubrication conditions, changes on only the dynamic properties of the experimental set up (e.g. stiffness) or on the driving velocity, produce dramatic changes in the observed characteristics.

It thus appears that these experimental characteristics are not at all an intrinsic property of the pairs of surfaces in contact. Such a point of view has been expressed by several authors who attribute apparent decreases of the coefficient of friction with the increase in sliding velocity to the occurrence of micro-vibrations of the slider in the direction normal to the contact surface. In fact, Tolstoi observed in (ref.4) that the forward movements of a slider during stick-slip motion invariably occur in strict syncronism with upward jumps. Oscillograms of the normal contact oscillations which occur during the sliding portions of the stick-slip cycles were obtained by Tolstoi, Borisova and Grigorova (ref.5). An examination of these oscillations revealed a fundamental frequency consistent with the normal interface stiffness properties (see Budanov, Kudinov and Tolstoi (ref.6)). Especially interesting is the experimental observation of Tolstoi (ref.4) that external normal damping of the normal oscillations of a slider can eliminate its stick-slip motion, and produce a smooth sliding with no quantitative difference between the so-called static and kinetic coefficients of friction.

These observations (and several others described in (ref.1)) led us to the conclusion that *an appropriate model for sliding friction must incorporate physically reasonable normal contact conditions*. A constitutive equation for the normal deformability of the interface and the Coulomb's law of friction will be thus the essential ingredients of the interface model to be developed in the following.

A model of interface response

Consider a continuous material body B, in contact with another material body B_1 over a contact surface $\Gamma_C \subset \partial B$.

The contact surface Γ_C represents the boundary of the parent bulk material of which the body B is composed. One can regard it as parallel to a surface representing the average surface height of the asperities of the physical body B. We suppose that Γ_C has a well defined exterior normal vector $\underset{\sim}{n}$.

For simplicity of presentation, but with easy generalization, we assume that the body B_1 is rigid and ideally flat. In the spirit of Fig. 1b, we also assume that the body B_1 does not move on the direction of $\underset{\sim}{n}$, but that it can move with some prescribed velocity $\dot{\underset{\sim}{U}}{}_T^C$ parallel to Γ_C.

We suppose that the actual interface (asperities, oxide film, adsorbed gas, work-hardened material, etc.) is initially of thickness t_o as shown in Fig. 3. The initial gap g between B and B_1 is defined as the distance, along the direction of the normal vector $\underset{\sim}{n}$, between the highest asperities of the body B and the flat surface of B_1 in the reference (undeformed) configuration. The interface thickness after deformation is denoted by t in Fig. 3, and the actual displacement of Γ_C in the direction of $\underset{\sim}{n}$ is $u_n = \underset{\sim}{u} \cdot \underset{\sim}{n}$. Thus, the approach of the material contact surfaces is $a = t_o - t = (u_n - g)_+$ where $(\cdot)_+ = \max\{0,\cdot\}$.

On the other hand, if $\dot{\underset{\sim}{u}}_T = \dot{\underset{\sim}{u}} - \dot{u}_n \underset{\sim}{n}$ denotes the tangential velocity of the points on Γ_C, then the relative sliding velocity between bodies B and B_1 is equal to $\dot{\underset{\sim}{u}}_T - \dot{\underset{\sim}{U}}{}_T^C$. Here $(\dot{\cdot})$ denotes partial differentiation with respect to time $\frac{\partial}{\partial t}(\ )$.

Denoting by σ_n and $\underset{\sim}{\sigma}_T$ the normal and tangential (frictional) stresses on Γ_C, respectively, the constitutive relations for the interface adopted here are the following:

Normal interface response

$$-\sigma_n = c_n (u_n - g)_+^{m_n} + b_n (u_n - g)_+^{\ell_n} \dot{u}_n \tag{1}$$

Friction conditions

$$u_n \leqq g \Longrightarrow \underset{\sim}{\sigma}_T = \underset{\sim}{0}$$

$$u_n > g \Longrightarrow \begin{cases} |\underset{\sim}{\sigma}_T| \leqq c_T (u_n - g)_+^{m_T} \\ \text{and} \\ |\underset{\sim}{\sigma}_T| < c_T (u_n - g)_+^{m_T} \Longrightarrow \dot{\underset{\sim}{u}}_T - \dot{\underset{\sim}{U}}{}_T^C = \underset{\sim}{0} \\ \text{and} \\ |\underset{\sim}{\sigma}_T| = c_T (u_n - g)_+^{m_T} \Longrightarrow \exists\, \lambda \geqq 0,\ \dot{\underset{\sim}{u}}_T - \dot{\underset{\sim}{U}}{}_T^C = -\lambda \underset{\sim}{\sigma}_T \end{cases} \tag{2}$$

Here c_n, m_n, b_n, ℓ_n, c_T, m_T are material parameters characterizing the interface and are to be determined experimentally.

The following remarks provide an explanation and interpretation of these relations:

1. The interface constitutive equation (1) combines a nonlinear power-law elastic contribution $-\sigma_n^e = c_n (u_n - g)_+^{m_n}$ with a nonlinear dissipative component given by $-\sigma_n^d = b_n (u_n - g)_+^{\ell_n} \dot{u}_n$. The form of the nonlinearly elastic contribution σ_n^e is consistent with experimental observations summarized in (ref.1,7) for the case of interfaces subjected to low nominal pressures, characteristic of sliding interfaces:

$$\left.\frac{d\sigma_n^e}{da}\right|_{a=0} = 0 \quad (a = (u_n - g)_+) \qquad\qquad -\sigma_n^e \propto a^{m_n} \text{ with } 2 \leq m_n \leq 3.33.$$

Tables with experimental values of the constants c_n and m_n for several combinations of materials and surface finishes can be found in (ref.7).

We also observe that, for metallic surfaces, after a period of rubbing and smoothing of the interface, its normal response is essentially elastic, if severe wear is prevented during the process of sliding (ref.1).

2. The nonlinear dissipative term σ_n^d is designed to model, only in an approximate manner, the hysteresis loops which are known to result from the actual elasto-plastic behavior of the interface asperities. Indeed, the constitutive equation (1) allows for the approximation of the loading paths of the form presented in Fig. 4a, which are known to occur if the surfaces are allowed to unload completely (Thornley et al. (ref.8)) by loops of the form presented in Fig. 4b. The idea of a similar approximation was proposed by Hunt and Crossley (ref.9) for vibroimpact phenomena involving macroscopic Hertzian contacts. For small energy losses, the correlation between the damping coefficient b_n and the energy loss per cycle of contact is readily obtainable (ref.9).

3. The friction law (2) is a slight generalization of Coulomb's friction law, which is recovered if $m_n = m_T$ and normal dissipative effects are negligible ($\sigma_n^d \approx 0$). In that case, the maximum value of the modulus of the friction stress is equal to the product of $\mu (= c_T/c_n)$ and the normal pressure ($|\underset{\sim}{\sigma}_T^{max}| = \mu|\sigma_n|$), where μ is the usual coefficient of friction. The law (2) allows for a dependence of the coefficient of friction on the normal contact pressure. The general form for this dependence, consistent with (2) (again for negligible σ_n^d), is the following:

$$\mu = C|\sigma_n|^{\alpha} \text{ with } \alpha = (m_T/m_n) - 1 \text{ and } C = c_T/c_n^{m_T/m_n}$$

4. In using the friction law (2), *we assume that no distinction between*

coefficients of static and kinetic friction exists. Here we follow, with a very simple model, the essential ideas of Tolstoi and co-workers (ref.4,5,6). In the next sections, we show that apparent decreases of measurable coefficients of friction can be the result of dynamic instabilities which are a consequence of the inherent non-symmetry of operators governing problems with friction.

SOME FRICTIONAL CONTACT PROBLEMS IN ELASTICITY THEORY

A problem in elastodynamics

We begin by considering a metallic body (Fig. 5) the interior of which is an open bounded domain in $\mathbb{R}^N$ (N=2 or 3). Points (particles) in Ω with cartesian coordinates x_i, $1 \leq i \leq N$, relative to a fixed coordinate frame are denoted by $\underset{\sim}{x}$ (= $(x_1\ x_2,\dots,\ x_N)$) and the volume measure by dx (= $dx_1\ dx_2 \dots dx_n$). The smooth boundary Γ of Ω contains four open subsets Γ_D, Γ_F, Γ_E, Γ_C, such that

$$\Gamma = \underset{\alpha}{\cup}\bar{\Gamma}_\alpha, \Gamma_\alpha \cap \Gamma_\beta = \emptyset \qquad \forall\alpha \neq \beta,\ \alpha,\beta \in \{D,F,E,C\} \tag{3}$$

and we denote by Σ the following open subset of Γ

$$\Sigma = \mathrm{int}(\Gamma-\Gamma_D) \tag{4}$$

Points on Γ with cartesian coordinates s_i, $1 \leq i \leq N$, are denoted by $\underset{\sim}{s}$ and the surface measure is denoted by ds.

We assume that the metallic body has a linearly elastic behavior characterized by the generalized Hooke's law,

$$\sigma_{ij}(\underset{\sim}{u}) = E_{ijk\ell}\, u_{k,\ell},\ 1 \leq i,\ j,\ k,\ \ell \leq N \tag{5}$$

where the σ_{ij} denote the components of the Cauchy stress tensor, $\underset{\sim}{u}$ = $(u_1,\ u_2,\dots,\ u_N)$ = $\underset{\sim}{u}(\underset{\sim}{x},t)$ is the vector field of displacements, the components of which are sufficiently smooth functions of position ($\underset{\sim}{x}$) and time (t) and $E_{ijk\ell} = E_{ijk\ell}(\underset{\sim}{x})$ are the usual elasticities of the material. In (5) and throughout this work $(\ )_{,\ell}$ denotes partial differentiation with respect to x_ℓ and the summation convention is used.

We suppose that body forces with components of force per unit volume b_i = $b_i\ (\underset{\sim}{x},t)$, $1 \leq i \leq N$, act in the body. We denote by $\rho = \rho(\underset{\sim}{x})$ the mass density of the material of which the body is composed. We also suppose that the body is fixed on Γ_D and that tractions $t_i = t_i(\underset{\sim}{s},t)$, $1 \leq i \leq N$, are prescribed on Γ_F. The body may be elastically supported on Γ_E by linear springs of moduli $K_{ij} = K_{ij}(\underset{\sim}{s})$. The undeformed configuration of those springs corresponds to the displacements $U_i^E = U_i^E(\underset{\sim}{s})$ on Γ_E.

We also suppose that the body may come in contact along a (candidate) contact surface Γ_C with a rough foundation which slides by the material contact surface with a velocity $\dot{\underset{\sim}{U}}{}_T^C = \dot{\underset{\sim}{U}}{}_T^C(\underset{\sim}{s})$ tangent to Γ_C. We assume that g, c_n, c_T, b_n are given as sufficiently smooth functions of $\underset{\sim}{s}$, have the physical meaning mentioned earlier, and that the real numbers m_n, m_T, ℓ_n are also specified.

Assuming sufficient smoothness for all the functions involved, the equations governing this elastodynamics problem, for a time interval [0,T], are grouped as follows:

Linear momentum equations

$$\sigma_{ij}(\underset{\sim}{u})_{,j} + b_i = \rho \ddot{u}_i \text{ in } \Omega \times (0,T) \tag{6}$$

where the σ_{ij} satisfy the constitutive equations (5).

Boundary conditions

$$u_i = 0 \text{ on } \Gamma_D \times (0,T) \tag{7}$$

$$\sigma_{ij}(\underset{\sim}{u})\, n_j = t_i \text{ on } \Gamma_F \times (0,T) \tag{8}$$

$$\sigma_{ij}(\underset{\sim}{u})\, n_j = -K_{ij}(u_j - U_j^E) \text{ on } \Gamma_E \times (0,T) \tag{9}$$

$$(1) \text{ and } (2) \text{ on } \Gamma_C \times (0,T) \tag{10}$$

Initial conditions

$$\left.\begin{aligned} \underset{\sim}{u}(\underset{\sim}{x},0) &= \bar{\underset{\sim}{u}}_o(\underset{\sim}{x}) \\ \dot{\underset{\sim}{u}}(\underset{\sim}{x},0) &= \bar{\underset{\sim}{u}}_1(x) \end{aligned}\right\} \quad \underset{\sim}{x} \in \Omega \tag{11}$$

A weak formulation for the elastodynamics problem above leads to a variational statement in the form of an inequality, which is a version of the principle of virtual power. Taking, for simplicity $\rho \equiv 1$, $b_n \equiv 0$, we have thus the following

Problem 1. Find a function $\underset{\sim}{u}$ such that

$$\langle \ddot{\underset{\sim}{u}}(t), \underset{\sim}{v}-\dot{\underset{\sim}{u}}(t) \rangle + a(\underset{\sim}{u}(t), \underset{\sim}{v}-\dot{\underset{\sim}{u}}(t)) + \langle P(u(t)), v-\dot{u}(t) \rangle + j(\underset{\sim}{u}(t),\underset{\sim}{v})$$
$$- j(\underset{\sim}{u}(t), \dot{\underset{\sim}{u}}(t)) \geqq \langle \underset{\sim}{f}(t), \underset{\sim}{v}-\dot{\underset{\sim}{u}}(t) \rangle \qquad \forall \underset{\sim}{v} \in V \tag{12}$$

with the initial conditions:

$$\underset{\sim}{u}(0) = \bar{\underset{\sim}{u}}_0 \qquad \dot{\underset{\sim}{u}}(0) = \bar{\underset{\sim}{u}}_1 \tag{13}$$

In the above $V = \{\underset{\sim}{v} \in (H^1(\Omega))^N | \gamma(\underset{\sim}{v}) = \underset{\sim}{0} \text{ a.e. on } \Gamma_D\}$ denotes the space of admissable displacements (velocities) and γ denotes the trace operator (ref. 10); $\langle \cdot,\cdot \rangle$ denotes duality pairing on V'xV where V' is the dual space of V;

$a(\underset{\sim}{u},\underset{\sim}{v}) = a_o(\underset{\sim}{u},\underset{\sim}{v}) + a_E(\gamma(\underset{\sim}{u}),\gamma(\underset{\sim}{v}))$ denotes the virtual work (power) produced by the deformation of the linearly elastic body Ω plus the linear springs on Γ_E; $\langle P(\underset{\sim}{u}),\underset{\sim}{v}\rangle$ denotes the virtual work(power) produced by the normal deformation of the contact interface; $j(\underset{\sim}{u},\underset{\sim}{v})$ denotes the virtual power produced by the frictional sliding and $\langle f(t),\underset{\sim}{v}\rangle$ denotes the virtual work (power) produced by the external forces on the displacement (velocity) $\underset{\sim}{v}$. Assuming that the data satisfy suitable smoothness requirements, the definitions of these forms are the following

$$a_o(\underset{\sim}{u},\underset{\sim}{v}) = \int_\Omega E_{ijk\ell}\, u_{k,\ell}\, v_{i,j}\; dx \qquad \underset{\sim}{u},\underset{\sim}{v} \in V$$

$$a_E(\underset{\sim}{\xi},\underset{\sim}{\eta}) = \int_{\Gamma_E} K_{ij}\, \xi_j\, n_i\; ds \qquad \underset{\sim}{\xi},\underset{\sim}{\eta} \in \gamma(V)$$

$$\langle P(\underset{\sim}{u}),\underset{\sim}{v}\rangle = \int_{\Gamma_C} c_n(u_n-g)_+^{m_n}\, v_n ds \qquad \underset{\sim}{u},\underset{\sim}{v} \in V \tag{14}$$

$$j(\underset{\sim}{u},\underset{\sim}{v}) = \int_{\Gamma_C} c_T(u_n-g)_+^{m_T}\, |v_T-\dot{U}_T^C| ds \qquad \underset{\sim}{u},\underset{\sim}{v} \in V$$

$$\langle f(t),\underset{\sim}{v}\rangle = \int_\Omega \underset{\sim}{b}(t)\cdot \underset{\sim}{v} dx + \int_{\Gamma_F} \underset{\sim}{t}(t)\cdot\gamma(\underset{\sim}{v}) ds + \int_{\Gamma_E} K_{ij}U_j^E v_i ds \qquad \underset{\sim}{v} \in V$$

We now turn our attention to the approximation of Problem 1 by a family of regularized problems more suitable for computational purposes.

Toward this end, we approximate the friction functional $j: V \times V \to \mathbb{R}$, which is nondifferentiable in the second argument (velocity), by a family of regularized functionals j_ϵ which are convex and differentiable in the second argument:

$$j_\epsilon(\underset{\sim}{u},\underset{\sim}{v}) = \int_{\Gamma_C} c_T(u_n-g)_+^{m_T}\psi_\epsilon(\underset{\sim}{v}_T-\dot{U}_T^C)\; ds \qquad \underset{\sim}{u},\underset{\sim}{v} \in V \tag{15}$$

Here the function $\psi_\epsilon: (L^q(\Gamma_C))^N \to L^q(\Gamma_C)$ is a continuously differentiable approximation of the function $|\cdot|: (L^q(\Gamma_C))^N \to L^q(\Gamma_C)$ (ref.1,11,12).

The partial derivative of j_ϵ with respect to the second argument, at (u,w) in the direction of v, is then given by:

$$\langle J_\epsilon(u,w), \underset{\sim}{v}\rangle \equiv \langle \partial_2 j_\epsilon(u,w), \underset{\sim}{v}\rangle$$

$$= \int_{\Gamma_C} c_T(u_n-g)_+^{m_T}\, [\phi_\epsilon(\underset{\sim}{w}_T-\dot{U}_T^C)\,(\underset{\sim}{v}_T)] ds \qquad u,v,w \in V \tag{16}$$

where $\phi_\epsilon(\underset{\sim}{\xi})(\underset{\sim}{\eta}) \equiv \psi'_\epsilon(\underset{\sim}{\xi})(\underset{\sim}{\eta})$, $\underset{\sim}{\xi},\underset{\sim}{\eta} \in (L^q(\Gamma_C))^N$, is the directional derivative of ψ_ϵ at $\underset{\sim}{\xi}$ on the direction of $\underset{\sim}{\eta}$.

We now define the regularized form of Problem 1:

Problem 1_ε. Find a function $\underset{\sim}{u}_\varepsilon$ such that

$$\langle \ddot{\underset{\sim}{u}}_\varepsilon(t), \underset{\sim}{v} \rangle + a(\underset{\sim}{u}_\varepsilon(t), \underset{\sim}{v}) + \langle \underset{\sim}{P}(\underset{\sim}{u}_\varepsilon(t)), \underset{\sim}{v} \rangle$$

$$+ \langle \underset{\sim}{J}_\varepsilon(\underset{\sim}{u}_\varepsilon(t), \dot{\underset{\sim}{u}}_\varepsilon(t)), \underset{\sim}{v} \rangle = \langle \underset{\sim}{f}(t), \underset{\sim}{v} \rangle \qquad \forall \underset{\sim}{v} \in V \tag{17}$$

with the initial conditions

$$\underset{\sim}{u}_\varepsilon(0) = \bar{\underset{\sim}{u}}_0 \qquad \dot{\underset{\sim}{u}}_\varepsilon(0) = \bar{\underset{\sim}{u}}_1 \quad . \tag{18}$$

<u>The steady frictional sliding of a metallic body</u>

We consider here a metallic body satisfying the same constitutive equation and submitted to the same type of forces and boundary conditions as in the previous section. The essential difference is that now we seek an equilibrium position of the body in unilateral contact with the rough moving foundation. Denoting by $\underset{\sim}{\tau} = \underset{\sim}{\tau}(\underset{\sim}{s})$ the unit tangent vector at each point of Γ_C parallel to the prescribed (non-zero) velocity $\dot{\underset{\sim}{U}}{}_T^C$ of the rough foundation, the friction conditions (2) simplify to the

Steady sliding friction equation:

$$\underset{\sim}{\sigma}_T = c_T(u_n - g)_+^{m_T} \underset{\sim}{\tau} \text{ on } \Gamma_C \tag{19}$$

This is the essential modification to be introduced in the governing equations (6-10). Obviously we have $\dot{\underset{\sim}{u}} = \ddot{\underset{\sim}{u}} = \underset{\sim}{0}$ and all the functions are assumed independent of time. Also, the domains $\Omega \times (0,T)$, $\Gamma_D \times (0,T)$, etc. in (6-10) should be replaced by Ω, Γ_D, etc.

Let us now assume that the domain Ω is such that

$$\Omega \in C^{0,1} \tag{20}$$

and that the data satisfy the following conditions (with $1 \leq i,j,k,\ell \leq N$)

$$E_{ijk\ell} \in L^\infty(\Omega);\ E_{ijk\ell} = E_{jik\ell} = E_{ij\ell k} = E_{k\ell ij} \text{ a.e. in } \Omega$$

$$\exists M_0 > 0 \text{ such that } \max_{1 \leq i,j,k,\ell \leq N} \|E_{ijk\ell}\|_{\infty,\Omega} \leq M_0 \tag{21}$$

$$\exists \alpha_0 > 0 \text{ such that } E_{ijk\ell}A_{k\ell}A_{ij} \geq \alpha_0 A_{ij}A_{ij} \text{ a.e. in } \Omega, \text{ for every symmetric array } A_{ij}$$

$$K_{ij} \in L^\infty(\Sigma);\ K_{ij} = K_{ji} \text{ a.e. on } \Sigma;\ K_{ij} = 0 \text{ a.e. on } \Gamma_F \cup \Gamma_C$$

$$\exists M_E > 0 \text{ such that } \max_{1 \leq i,j \leq N} \|K_{ij}\|_{\infty,\Sigma} \leq M_E \tag{22}$$

$$\exists \alpha_E > 0 \text{ such that } K_{ij}a_j a_i \geq \alpha_E a_i a_i \text{ a.e. on } \Gamma_E \text{ for every vector } a_i$$

$b_i \in L^2(\Omega)$; $t_i \in L^{q'}(\Sigma)$, $t_i = 0$ a.e. on $\Gamma_E \cup \Gamma_C$; $U_i^E \in L^{q'}(\Sigma)$;

$$\tau_i \in L^{\infty}(\Sigma),\ \sum_{i=1}^{N} \tau_i^2 = 1,\ \sum_{i=1}^{N} \tau_i n_i = 0 \text{ a.e. on } \Sigma;\ g \in L^q(\Sigma); \tag{23}$$

$c_n, c_T \in L^{\infty}(\Sigma)$, $c_n = c_T = 0$ a.e. on $\Gamma_F \cup \Gamma_E$

In the above q and q' satisfy

$$q = 1+m_0;\ m_0 = \max\{m_n, m_T\};\ 1/q+1/q' = 1 \tag{24}$$

where $m_n, m_T \in \mathbb{R}$ satisfy

$$1 < m_n, m_T \text{ if } N=2;\ 1 < m_n, m_T \leq 3 \text{ if } N=3 \tag{25}$$

The virtual work produced by the friction stresses on the displacement $\underset{\sim}{v}$ is now given by

$$\langle J(\underset{\sim}{u}), \underset{\sim}{v}\rangle \equiv \langle J_\varepsilon(\underset{\sim}{u}, \underset{\sim}{0}), \underset{\sim}{v}\rangle = -\int_{\Gamma_C} c_T (u_n - g)_+^{m_T} \underset{\sim}{\tau} \cdot \underset{\sim}{v}_T ds \qquad \underset{\sim}{u}, \underset{\sim}{v} \in V \tag{26}$$

It is then possible to show that the classical formulation of the <u>steady sliding problem</u> is equivalent, in an appropriate sense (ref.13), to the following variational formulation.

Problem 2. Find a function $\underset{\sim}{u}_0 \in V$ such that

$$a(\underset{\sim}{u}_0, \underset{\sim}{v}) + \langle P(\underset{\sim}{u}_0), \underset{\sim}{v}\rangle + \langle J(\underset{\sim}{u}_0), \underset{\sim}{v}\rangle = \langle \underset{\sim}{f}, \underset{\sim}{v}\rangle \qquad \forall \underset{\sim}{v} \in V \tag{27}$$

If the coefficient of friction <u>or</u> the applied forces and initial gap are "sufficiently small" it is also possible to prove (ref. 13) existence and uniqueness for the steady sliding problem:

Proposition. Let assumptions (3), (4) and (20) on Ω, Γ and Σ hold together with assumptions (21-25) on the data. Let meas $\Gamma_D > 0$ <u>or</u> meas $\Gamma_E > 0$. Let the following assumptions also hold:

$$\text{There exists } \underset{\sim}{w}^* \in V \text{ such that } \gamma(\underset{\sim}{w}^*) \cdot \underset{\sim}{n} = g \text{ a.e. } \Sigma \tag{28}$$

$$m_n, m_T < 3 \text{ if } N = 3 \tag{29}$$

Then there exists a constant $C = C(m_T, q, N, \Omega, \Sigma) > 0$, such that if, in addition,

$$\frac{\|c_T\|_{\infty, \Gamma_C} (\|\underset{\sim}{f}\|_{V'} + M\|\underset{\sim}{w}^*\|_V)^{m_T - 1}}{\alpha^{m_T}} \leq C \tag{30}$$

with M and α denoting the continuity and coercivity constants for the bilinear form $a(\cdot,\cdot)$, then there exists a $R > 0$ such that Problem 2 has a unique solution in the closed ball $K = \{\underset{\sim}{v} \in V \mid \|\underset{\sim}{v} - \underset{\sim}{w}^*\|_V \leq R\}$.

Dynamic stability of the steady frictional sliding

We are interested here in the analysis of the dynamic stability of the steady sliding equilibrium position $\underset{\sim}{u}_0$. We restrict ourselves to two dimensional problems (N=2).

A natural idea is to study the behavior of a linearized version of (12,17) in the neighborhood of $\underset{\sim}{u}_0$. Such linearization is possible because the nonlinear operator

$A : V \longrightarrow V'$

$$\langle A(\underset{\sim}{u}),\underset{\sim}{v}\rangle = a(\underset{\sim}{u},\underset{\sim}{v}) + \langle P(\underset{\sim}{u}),\underset{\sim}{v}\rangle + \langle J(\underset{\sim}{u}),\underset{\sim}{v}\rangle \qquad \underset{\sim}{u},\underset{\sim}{v} \in V$$

emerging from (27) and in (12,17) for $\dot{\underset{\sim}{u}}_T(\underset{\sim}{s}) = \underset{\sim}{0}$, $|\dot{\underset{\sim}{U}}{}^C_T(\underset{\sim}{s})| > \varepsilon > 0$ a.e. $\underset{\sim}{s} \in \Gamma_C$, is continuously differentiable in V (ref.13).

A first step toward the stability analysis of the equilibrium position $\underset{\sim}{u}_0$ is then the formulation of the eigenvalue problem associated with the sesquilinear form $\tilde{a}(\cdot,\cdot)$ defined by the derivative of A at $\underset{\sim}{u} = \underset{\sim}{u}_0$ (fixed):

$$\tilde{a}(\underset{\sim}{w},\underset{\sim}{v}) = \int_\Omega E_{ijk\ell} w_{k,\ell} \bar{v}_{i,j} dx + \int_{\Gamma_E} K_{ij} w_j \bar{v}_i ds + \int_{\Gamma_C} K_n w_n \bar{v}_n ds + \int_{\Gamma_C} K_{Tn} w_n \bar{v}_\tau ds$$

where superimposed bars denote complex conjugation, $v_\tau = \underset{\sim}{v}_T \cdot \underset{\sim}{\tau} = \gamma(\underset{\sim}{v}) \cdot \underset{\sim}{\tau}$, K_n denotes the linearized normal stiffness of the foundation at the equilibruim position $\underset{\sim}{u}_0$ and K_{Tn} denotes the coupling stiffness coefficient between normal displacements and tangential stresses, also at the equilibruim position $\underset{\sim}{u}_0$, i.e.,

$$K_n = m_n c_n (u_{on} - g)_+^{m_n - 1} \qquad K_{Tn} = -m_T c_T (u_{on} - g)_+^{m_T - 1} .$$

The eigenvalue problem associated with the non-symmetric form $\tilde{a}(\cdot,\cdot)$ is then:

Problem 3. Find the values $\lambda \in \mathbb{C}$ for which there exists $\underset{\sim}{w} \in V$, $\underset{\sim}{w} \neq \underset{\sim}{0}$, such that

$$\tilde{a}(\underset{\sim}{w},\underset{\sim}{v}) + \lambda^2 (\underset{\sim}{w},\underset{\sim}{v}) = 0 \qquad \forall \underset{\sim}{v} \in V \tag{31}$$

Here, the L^2-innerproduct is $(\underset{\sim}{u},\underset{\sim}{v}) = \int_\Omega u_i \bar{v}_i dx$, $\underset{\sim}{u},\underset{\sim}{v} \in (L^2(\Omega))^N$.

FINITE ELEMENT APPROXIMATIONS AND ALGORITHMS

The discrete problems

Using standard finite element procedures, approximate versions of Problems 1_ε, 2 and 3 can be constructed in finite dimensional subspaces $V_h (\subset V \subset V')$. Here, we shall restrict ourselves to present the form of the resulting systems of equations for those discrete problems.

The finite elements approximation of Problem 1_ε leads to the following system of nonlinear ordinary differential equations,

$$\underset{\sim}{M}\,\ddot{\underset{\sim}{r}}(t) + \underset{\sim}{K}\,\underset{\sim}{r}(t) + \underset{\sim}{P}(\underset{\sim}{r}(t)) + \underset{\sim}{J}(\underset{\sim}{r}(t),\dot{\underset{\sim}{r}}(t)) = \underset{\sim}{F}(t) \tag{32}$$

with the initial conditions,

$$\underset{\sim}{r}(0) = \bar{\underset{\sim}{r}}_0; \qquad \dot{\underset{\sim}{r}}(0) = \bar{\underset{\sim}{r}}_1 \tag{33}$$

Here $\underset{\sim}{r}(t)$, $\dot{\underset{\sim}{r}}(t)$ and $\ddot{\underset{\sim}{r}}(t)$ denote the column vectors of nodal displacements, velocities and accelerations, respectively; $\underset{\sim}{M}(\underset{\sim}{K})$ is the standard mass (stiffness) matrix; $\underset{\sim}{F}(t)$ the vector of consistent nodal forces; $\underset{\sim}{P}(\underset{\sim}{r})$ and $\underset{\sim}{J}(\underset{\sim}{r},\dot{\underset{\sim}{r}})$ are the vectors of nodal normal and friction forces on Γ_C and $\bar{\underset{\sim}{r}}_0(\bar{\underset{\sim}{r}}_1)$ is the vector of initial nodal displacements (velocities).

Using the same notations above, the steady sliding Problem 2 leads to the following system of nonlinear algebraic equations (set $\dot{\underset{\sim}{r}} = \ddot{\underset{\sim}{r}} = \underset{\sim}{0}$ in (32)),

$$\underset{\sim}{K}\,\underset{\sim}{r}_0 + P(\underset{\sim}{r}_0) + \underset{\sim}{J}(\underset{\sim}{r}_0,\underset{\sim}{0}) = \underset{\sim}{F} \tag{34}$$

and Problem 3 leads to the following algebraic eigenvalue problem

$$[\tilde{\underset{\sim}{K}} + \lambda^2\,\underset{\sim}{M}]\,\underset{\sim}{w} = \underset{\sim}{0}, \quad \lambda \in \mathbb{C}, \quad \underset{\sim}{w} \neq \underset{\sim}{0} \tag{35}$$

where

$$\tilde{\underset{\sim}{K}} \equiv \underset{\sim}{K} + \underset{\sim}{K}^n(\underset{\sim}{r}_0) + \underset{\sim}{K}^{Tn}(\underset{\sim}{r}_0,\underset{\sim}{0})$$

$$\underset{\sim}{K}^n(\underset{\sim}{r}) \overset{\text{def}}{=} \frac{\partial}{\partial \underset{\sim}{r}} P(\underset{\sim}{r})$$

$$\underset{\sim}{K}^{Tn}(\underset{\sim}{r},\dot{\underset{\sim}{r}}) \overset{\text{def}}{=} \frac{\partial}{\partial \underset{\sim}{r}} J(\underset{\sim}{r},\dot{\underset{\sim}{r}})$$

We stress the fact that the matrix $\underset{\sim}{K}^{Tn}$ is not symmetric. This results from the inherent non-symmetry of the coupling between normal and tangential variables on the interface: normal stresses depend only upon the magnitude of the normal penetration, but the friction stresses depend not only upon the tangential velocities but also upon the normal penetration. It is precisely this non-symmetry that makes the eigenvalue problem (35) different from standard structural dynamics eigenvalue problems. Effects of this lack of symmetry on the dynamic response of the elastic body are discussed in the numerical examples.

Finally we observe that dissipative effects other than dry friction (linear viscous terms, normal contact damping of the form (1)) can be easily added to the developments of the previous sections. The major consequence for the eigenvalue problem (35) is then the introduction of additional terms, linear in λ.

Algorithms

The algorithms that we have used for solving the discrete dynamic system (32) involve variants of standard schemes in use in nonlinear structural mechanics calculations: Newmark's method and explicit central-difference technique, both associated with Newton-Raphson iterations within each time step. For further details see (ref.1).

In order to compute the solutions of (34) for a certain range of values of c_T, we subdivide the interval $[0,\bar{c}_T]$ into a prescribed number NINCT of increments $\Delta c_T = \bar{c}_T/\text{NINCT}$ and, for the Kth increment, K=0,...,NINCT, we again use the Newton-Raphson method to solve the nonlinear system (34) at each value of c_T.

For the computed equilibrium position at each increment K, the nonsymmetric eigenvalue problem (35) is solved using standard eigenvalue routines.

NUMERICAL EXAMPLES

The steady sliding of a block on a moving foundation

We consider here a block sliding, with friction, on a moving foundation (see Fig. 6). We assume that the block has a linearly elastic behavior with a Young's modulus $E = 1.4 \times 10^6 (10^3\text{Kg cm}^{-1}\text{ s}^{-2})$ and a Poisson's ratio $\nu = 0.25$. For simplicity, we assume that the body is in a state of plane strain. The geometry, total mass (M), total weight (W), total tangential stiffness (K_x) and contact properties are given as follows:

$$\left.\begin{array}{ll} L = 48.8\text{ cm} & M = 450\text{ Kg} \\ H = 30.5\text{ cm} & W = 450\ 10^3\text{Kg cm s}^{-2} \\ c_n = 10^{10}\ 10^3\text{Kg cm}^{-3.5}\text{ s}^{-2}, & m_n = m_T = 2.5,\ c_T = \mu c_n \end{array}\right\} \qquad (36)$$

$$B_0 = 30.5\text{ cm} \qquad K_x = 2388\quad 10^3\text{Kg s}^{-2} \qquad \mu \in [0,1.6) \qquad (37)$$

The finite elements model consists of a 4x3 mesh of nine-node isoparametric elements as depicted in Fig. 7.

A simple calculation reveals that for the given geometry a necessary condition for equilibrium is that $\mu < L/H = 1.6$; for $\mu = 1.6$ the body will begin to tumble (with unbounded rotations). In this section we compute the steady sliding equilibrium positions of the block for several values of μ in the admissable range $[0,1.6)$ and for each of those configurations we solve the eigenvalue problem (35).

In Fig. 7, we show the deformed mesh configurations at several values of μ. The most important information however comes from the corresponding eigenvalues. In the absence of damping, all the eigenvalues are pure imaginary for small values of μ. For $\mu \geq 0.32$ (see Fig. 8) the occurrence of eigenvalues with positive real part is observed. This implies that for such values of μ the steady sliding is dynamically unstable.

As might be expected, for the small magnitude of the applied forces, the block behaves much like a rigid body (see Fig. 7). For this reason, we have performed also the same computations, for the same block of Fig. 6 and with the same data (36,37), assuming that the block is a rigid body in plane motion. The corresponding three degrees of freedom are indicated in Fig. 6: the sliding and penetrating (normal) displacements of the center of mass, u_{xG} and u_{yG}, and the rotation, u_{Θ}.

The eigenvalues associated with the tangential displacement are a conjugate imaginary pair ($\pm i\sqrt{K_x/M}$) which does not vary with μ. In Fig. 9 we show the evolution of the four eigenvalues of (37) associated with the normal and rotational freedoms. As expected all of these eigenvalues compare favorably with the eigenvalues associated with similar modes obtained with the finite elements model (ref.1).

Perturbation of the steady sliding of a block on a moving foundation

In this section we show the effects of the eigenvalue structure of (35) on the motion of the block of Fig. 6. For each set of data (in particular some value of μ) we solve numerically the nonlinear equations of motion (32) with the following initial conditions: the initial displacements are prescribed as those corresponding to the steady sliding equilibrium configuration appropriate for the value of μ considered; initial velocities represent a small perturbation (upwards) of the translational velocity ($\dot{u}_{xG} = 0.0$, $\dot{u}_{yG} = -0.01\text{cm s}^{-1}$, $\dot{u}_{\Theta} = 0.0$). Various cases were run assuming either a linearly elastic block (the finite element model of Fig. 7) or a rigid block (the three degrees of freedom model of Fig. 6). Other than friction, the only dissipative effect considered with the first model was a normal contact dissipation of the form in (1) with $b_n = 0.381 \times 10^{10}$ ($10^3\text{Kg}\cdot\text{cm}^{-4} \cdot \text{s}^{-1}$) and $\ell_n = 2.5$; with the second model the results shown here were obtained with linear viscous damping characterized by diagonal coefficients C_x, C_y, C_{Θ}. Other data is given in the figures. The finite element results were obtained using the central-difference algorithm with $\Delta t_{max} = 3 \times 10^{-6}$ s and a diagonalized mass matrix. The rigid body motions were calculated using Newmark's method with $\Delta t_{max} = 10^{-5}$ s.

We first consider the small friction case, i.e., when (in the absence of damping) the eigenvalues λ of (35) are all imaginary. In this case, starting from the initial conditions mentioned above an apparently stable small-amplitude oscillation is obtained (ref.1).

More interesting facts occur if μ is sufficiently large that some of the eigenvalues of (35) have (even in the presence of some damping) positive real parts. The following remarks summarize our interpretation of the numerical results in this situation:

Due to the instability associated with normal and rotational modes, the normal and rotational oscillations grow (Fig. 10).

The variation of the normal force on the contact produces changes in the sliding friction force which in turn produce a tangential oscillation.

The tangential oscillation may then become sufficiently large that, for small values of the belt velocity $\dot{U}_x^C$ the points of the body on the contact surface attain the velocity $\dot{U}_x^C$ and the body sticks for short intervals of time (see Figs. 11, 15).

With the increase in magnitude of the normal oscillations, actual normal jumps of the body may occur (see Figs. 10, 14).

The repeated periods of adhesion have the result of decreasing the average value of the friction force on the contact and, due to the absence of equilibrium with the restoring force on the tangential spring, the tangential displacement of the center of mass decreases (see Figs. 12, 13).

It follows that for given geometry and material data, one of two following situations may occur: (a) for values of $\dot{U}_x^C$ larger than some critical value, the normal, rotational and tangential oscillations evolve to what appears to be a steady oscillation with successive periods of adhesion and sliding, the average values of the friction force and of the spring elongation being smaller than those corresponding to the steady sliding equilibrium position (see Fig. 12, $\dot{U}_x^C = 0.287 \text{ cm s}^{-1}$). (b) for values of $\dot{U}_x^C$ lower than the critical value, and at a sufficiently small value of the spring elongation, the normal (and rotational) damping is able to damp out the normal (and rotational) oscillation and the body sticks (see Figs. 12, 13) since the restoring force of the spring is then smaller than the maximum available friction force.

Thus, monitoring the spring elongations, as is often done in friction experiments, case (a) would be perceived as an apparently smooth sliding with a coefficient of kinetic friction smaller than the coefficient of static friction and case (b) would be perceived as stick-slip motion.

Acknowledgement. The authors gratefully acknowledge support of this work by the Air Force Office of Scientific Research under Contract F4950-84-0024.

REFERENCES

1 J.T. Oden and J.A.C. Martins, "Models and Computational Methods for Dynamic Friction Phenomena," Computer Methods in Appli. Mech. and Eng. (to appear)

2 F.P. Bowden and L. Leben, "The Nature of Sliding and the Analysis of Friction," Proc. Roy. Soc. Lond., A169, pp. 371-391, 1939.

3 H. Blok, "Fundamental Mechanical Aspects of Boundary Lubrication, S.A.E. Journal, 46(2), pp. 54-68, 1940.
4 D.M. Tolstoi, "Significance of the Normal Degree of Freedom and Natural Normal Vibrations Contact Friction, Wear, 10, pp. 199-213, 1967.
5 D.M. Tolstoi, G.A. Borisova and S.R. Grigorova, "Role of Intrinsic Contact Oscillations in Normal Direction During Friction," Nature of the Friction of Solids, Nauka i Tekhnika, p. 116, Minsk, 1971.
6 B.V. Budanov, V.A. Kudinov and D.M. Tolstoi, "Interaction of Friction and Vibration," Trenie i Iznos, vol. 1, No. 1, pp. 79-89, 1980.
7 N. Back, M. Burdekin and A. Cowley, "Review of the research on Fixed and Sliding Joints," Proc. 13th International Machine Tool, Des. and Res. Conf., ed. by S.A. Tobias and F. Koenigsberger, MacMillan, London, 1973.
8 R.H. Thornley, R. Connolly, M.M. Barash and F. Koenigsberger, "The Effect of Surface Topography Upon the Static Stiffness of Machine Tool Joints," Int. J. Mach. Tool Des. Res., 5, pp. 57-74, 1965.
9 K.H. Hunt and F.R.E. Crossley, "Coefficient of Restitution Interpreted as Damping in Vibroimpact," Journal of Applied Mechanics, pp. 440-445, June, 1975.
10 Adams, Sobolev Spaces, Academic Press, N.Y., 1977.
11 G. Duvaut and J.L. Lions, Inequalities in Mechanics and Physics Springer-Verlag, Berlin, Heidelberg, New York, 1976.
12 J.A.C. Martins and J.T. Oden, "A Numerical Analysis of a Class of Problems in Elastodynamics with Friction," Comput. Meths. Appl. Mech. Engrg., 40, pp. 327-360,1983.
13 J.A.C. Martins, Some Dynamic Frictional Contact Problems Involving Metallic Bodies, Ph.D. dissertation, University of Texas at Austin, 1985.

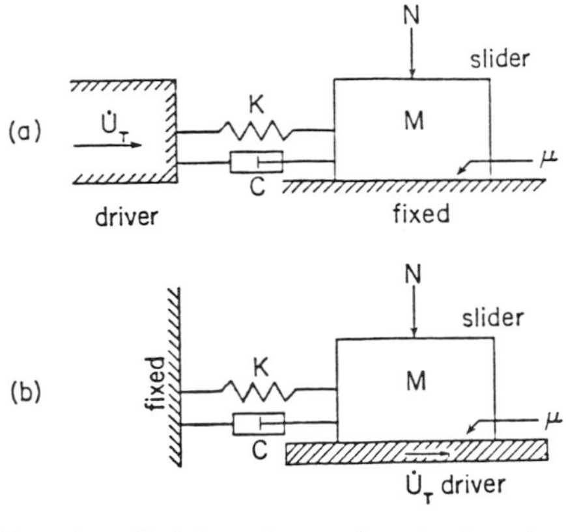

Fig. 1. Models of two (equivalent) sliding systems which may have stick-slip oscillations.

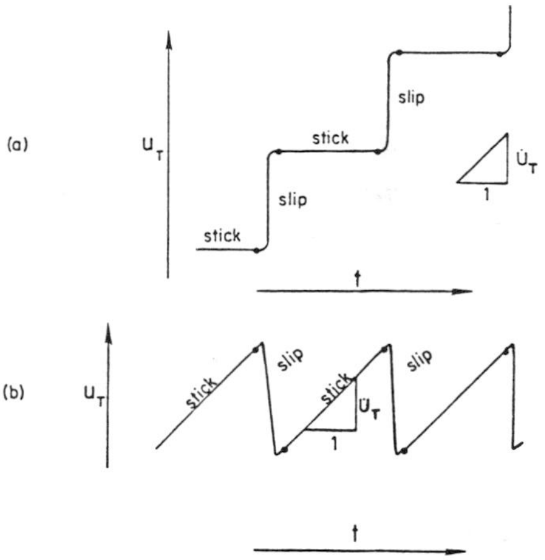

Fig. 2. Typical traces of stick-slip motion for systems (a) and (b) in Fig. 1.

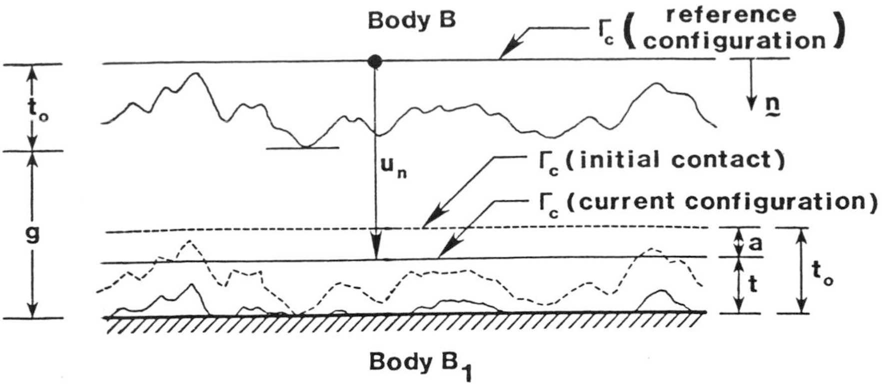

Fig. 3. Initial gap, normal displacement and penetrating approach at the contact surface Γ_C.

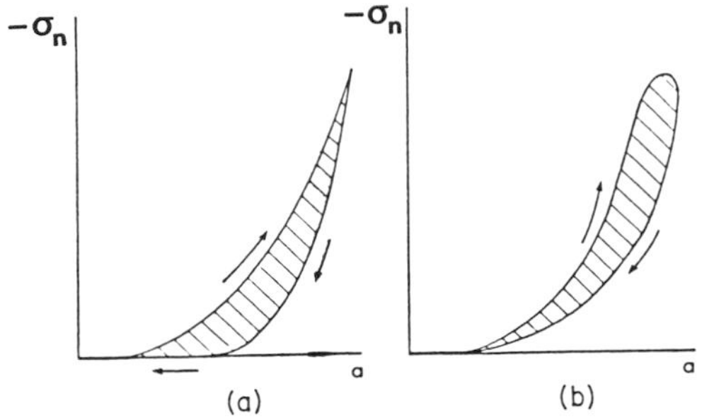

Fig. 4. Hysteresis loops for the normal deformation of the interface (schematic). (a) Experimentally observed loop, under quasi-static loading conditions. (b) Hysteresis loop modelled by the constitutive equation (1) under dynamic loading conditions.

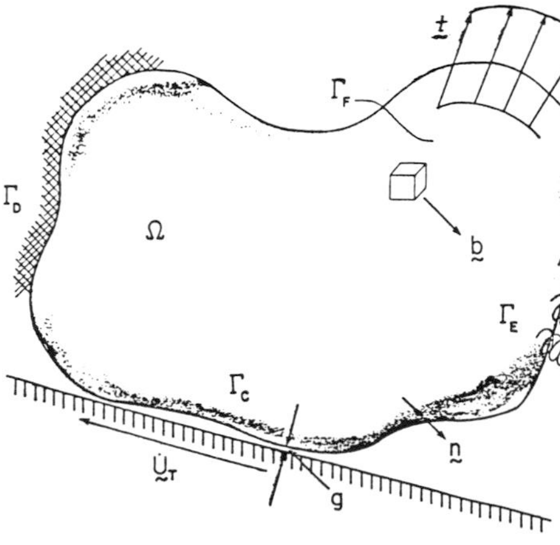

Fig. 5. Geometry and notation used in elastodynamics problem with sliding friction.

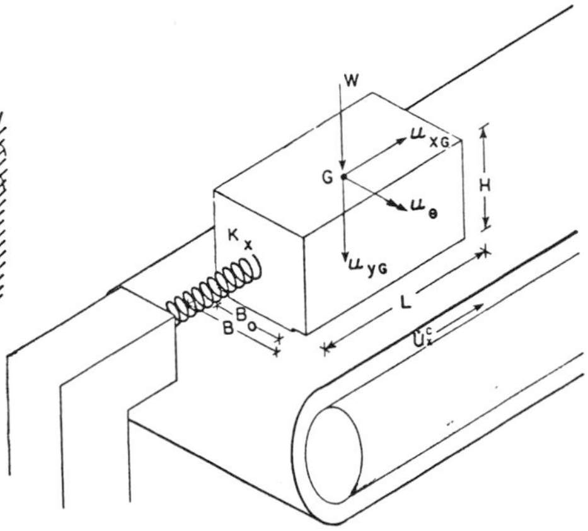

Fig. 6. A block sliding with friction on a moving belt.

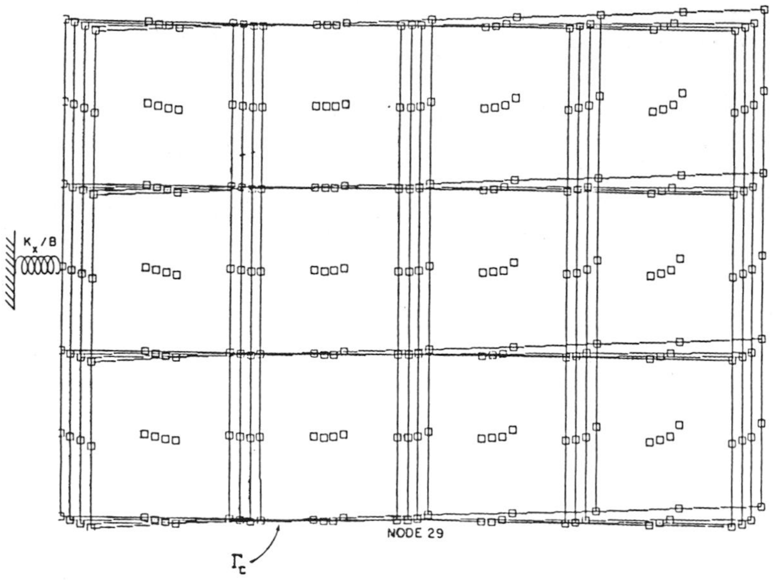

Fig. 7. Deformed configurations of a linearly elastic block for the steady sliding equilibrium configurations at several values of μ .

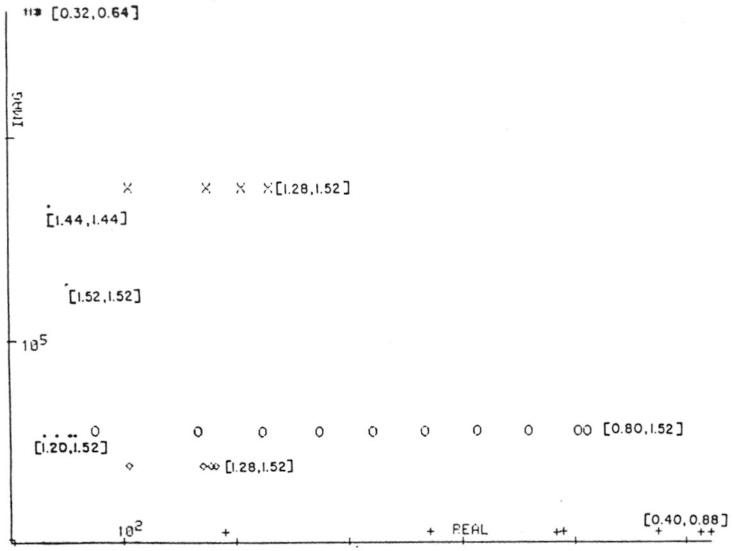

Fig. 8. Eigenvalues with positive real parts for a finite elements discretization of a block in steady sliding on a moving belt.

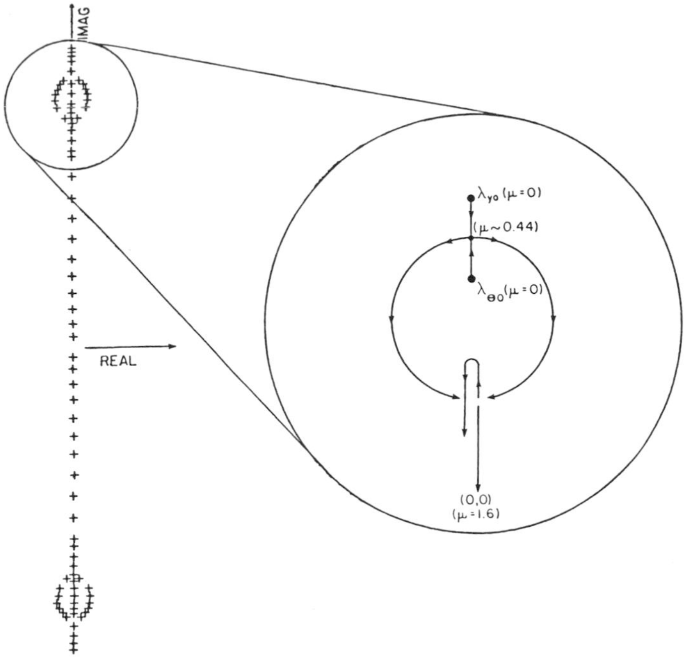

Fig. 9. Evolution of the Eigenvalues on the complex plane for the successive equilibrium positions obtained with increasing coefficient of friction.

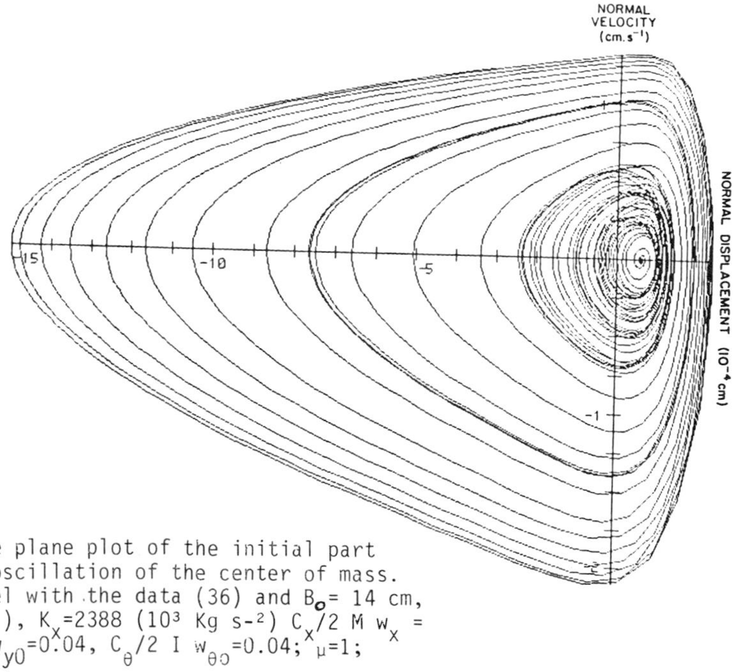

Fig. 10. Phase plane plot of the initial part of the normal oscillation of the center of mass. Rigid body model with the data (36) and B_o= 14 cm, I=70(10^3 Kg cm^2), K_x=2388 (10^3 Kg s^{-2}) $C_x/2\,M\,w_x$ = 0.005, $C_y/2\,M\,w_{y0}$=0.04, $C_\theta/2\,I\,w_{\theta 0}$=0.04; μ=1; $\dot{U}_x^C$=0.127 cm s^{-1}.

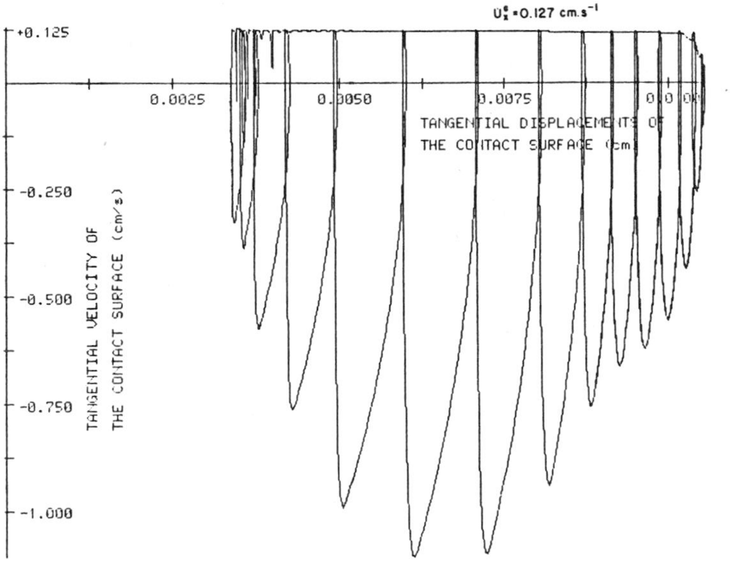

Fig. 11. Phase plane plot of the tangential motion of the points of the block on the contact surface: Rigid body model with the data of Fig. 10 except for K_x = 4441(10^3 Kg s^{-2}).

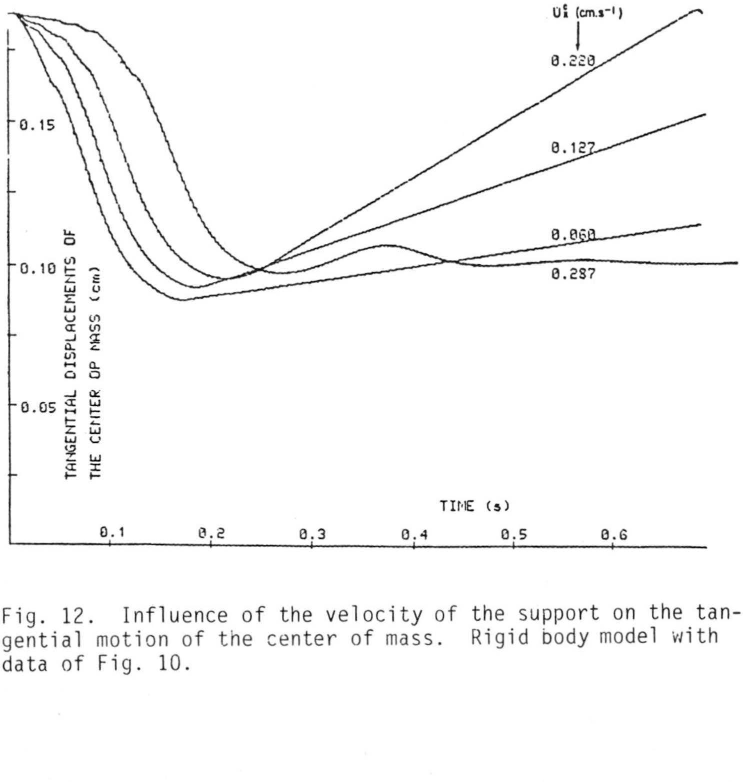

Fig. 12. Influence of the velocity of the support on the tangential motion of the center of mass. Rigid body model with data of Fig. 10.

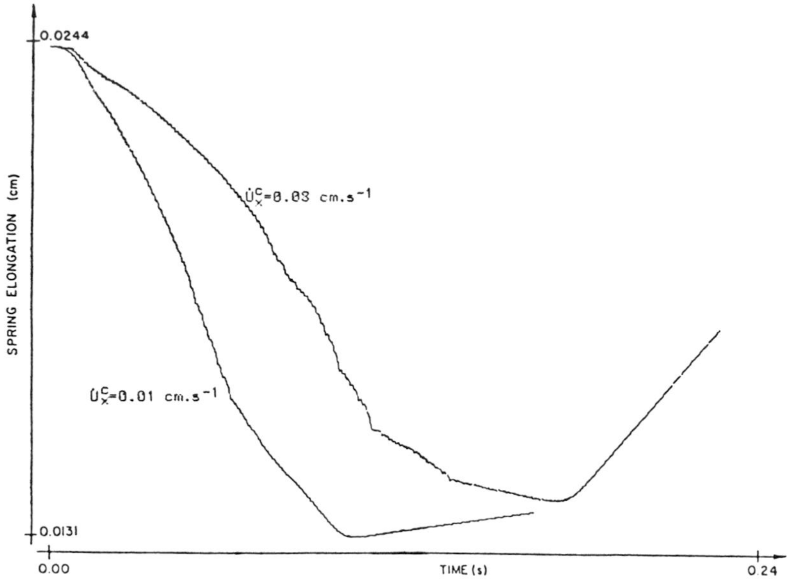

Fig. 13. Evolution of the spring elongation for two velocities $\dot{U}^C_x$. Finite element model with the data (36) and K_x = 11100 Kg s^{-2}, B_0=30.5 cm, μ= 0.6.

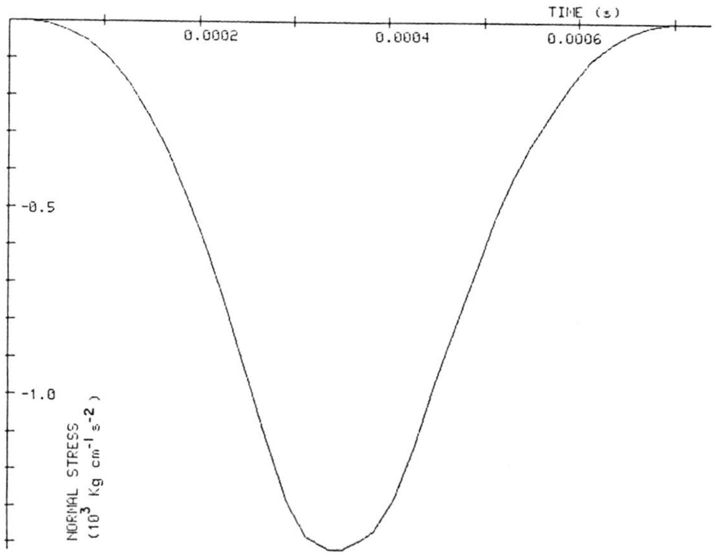

Fig. 14. Evolution of the normal contact stress at node 29 during the contact portion of one cycle of normal oscillation. Finite elements model with the data of Fig. 13 and $\dot{U}^C_x$ = 0.08 cm s^{-1}.

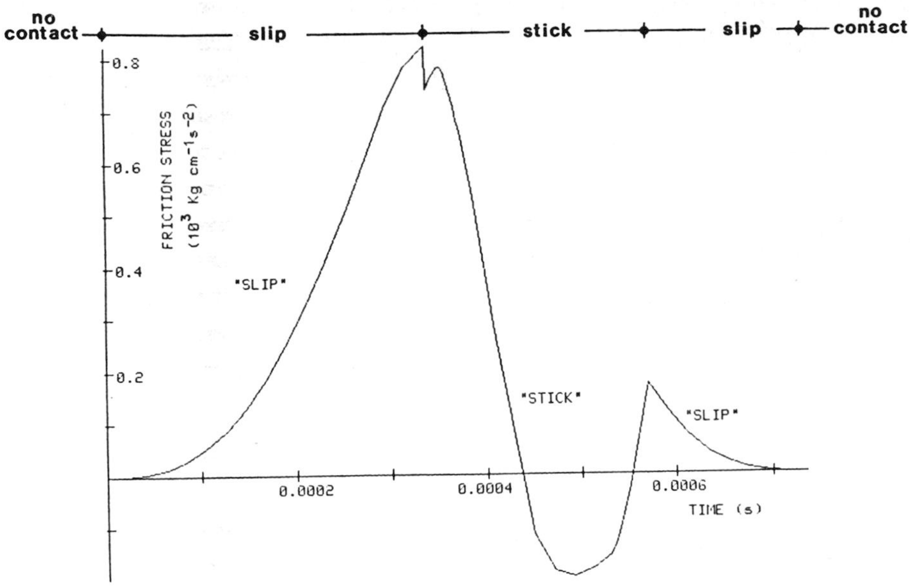

Fig. 15. Evolution of the friction stress at node 29 during the contact portion of one cycle of normal oscillation. Finite elements model with the data of Fig. 13 and $\dot{U}_x^c = 0.08$ cm s^{-1}.

HEMIVARIATIONAL INEQUALITIES IN FRICTIONAL CONTACT PROBLEMS AND APPLICATIONS

P. D. PANAGIOTOPOULOS

Department of Civil Engineering, Aristotle University Thessaloniki, Greece, and Institute for Technical Mechanics, Technical University of Aachen, Aachen, F.R.G.

ABSTRACT

The present paper deals with problems of nonmonotone frictional contact. For the boundary conditions describing the unilateral contact, nonconvex potential functions are considered. Several types of frictional conditions are examined. The respective boundary value problems have as variational formulations hemivariational inequalities which are formulated and studied. Certain algorithms for the numerical treatment of the arising problems are proposed and the theory is illustrated by numerical examples.

INTRODUCTION

A large number of contact problems in applied mechanics, formulated as variational inequalities have already been studied (see, e.g. [1] [2]). These inequalities express the principle of virtual work in its inequality form which is caused by the fact that the admissible variations of certain quantities are unilateral and this is the reason why we call such problems unilateral problems.

The first unilateral contact problem studied was the Signorini-Fichera problem [3] arising when a deformable body lies on a rigid support. Then we do not know a priori what part of the boundary remains in contact with the support and what part does not. If, moreover, frictional phenomena are taken into account another free boundary is to be determined between the parts with sliding friction and the ones with adhesive friction. It has been proved until now that the variational technique, characterized by the formulation and the solution of variational inequalities, offers a powerful tool for the treatment of such contact problems, especially if the nonsymmetry of the loading and the shape of the body does not permit the use of classical (bilateral) methods of numerical or analytical nature.

Numerous static and dynamic frictional contact problems of elastic structures with a deformable support or structure have been studied by many authors. For the respective literature we refer to [1] [2]. The frictional contact problems studied until now exhibit a common characteristic: the convexity of the potential of the contact and friction forces, thus giving rise to a monotone increasing relationship between reactions and displacements both in the normal and the tangential direction (cf. fig. 1). However, most practical problems are characterized by a nonmonotone displacement-reaction diagram, i.e. by nonconvex contact potentials, as, e.g. for the contact of a deformable body with a granular support. In these cases the lack of monotonicity does not permit the use of the classical techniques of variational inequalities based on monotonicity; then the arising variational forms are "hemivariational inequalities" [2], [4], [5].

NONMONOTONE BOUNDARY CONDITIONS OF UNILATERAL AND FRICTIONAL CONTACT

Nonmonotone Unilateral Contact

Let us denote by Ω the body under consideration and let Γ be its boundary. The body is referred to a fixed orthogonal Cartesian coordinate system $Ox_1x_2x_3$. We assume that the boundary Γ of the body is divided into three mutually disjoint parts Γ_U, Γ_F and Γ_S. On Γ_U the displacements $u=\{u_i\}$ have prescribed values

$$u_i = U_i \ , \ U_i = U_i(x) \tag{1}$$

and on Γ_F the boundary forces $S=\{S_i\}$ are given, i.e.

$$S_i = F_i \quad , \quad F_i = F_i(x). \tag{2}$$

Here $S_i = \sigma_{ij}n_j$ (i,j=1.2.3 - summation convention), where $\sigma=\{\sigma_{ij}\}$ is the stress tensor and $n=\{n_i\}$ is the outward unit normal vector to Γ. In order to formulate the conditions of unilateral contact, S (resp. u) is decomposed on Γ_S into normal and tangential components to Γ, say S_N and S_T (resp. u_N and u_T) and we assume generally a nonlinear relation between S_N and u_N (assumption of nonlinear Winkler springs), whereas in the tangential direction we assume for the present that S_T is given, i.e.

$$S_{Ti} = C_{Ti} \ , \quad C_{Ti} = C_{Ti}(x). \tag{3}$$

The contact between the body and the deformable support is idealised by a relation of the form:

$$\text{If} \quad u_N < 0 \quad \text{then} \quad S_N = 0 \qquad \text{on } \Gamma_S; \tag{4a}$$

$$\text{If} \quad u_N \geq 0 \quad \text{then} \quad S_N + k\,(u_N) = 0 \text{ on } \Gamma_S \tag{4b}$$

where $k(u_N)$ is (in contrast to [6]) a nonmonotone increasing function (Fig. 2a) which may also include jumps describing local crushing and locking effects. Function $k(u_N)$ may have different forms from point to point. If the contact surface may sustain a certain small tensile force (Fig. 2b) , we can write the following boundary condition:

$$\text{If} \quad u_N > \alpha \quad \text{then } S_N + k\,(u_N) = 0 \tag{5a}$$

if $u_N = \alpha$ then $0 \leq S_N \leq k(\alpha)$ (5b)

if $u_N < \alpha$ then $S_N = 0$ (5c)

If we had monotone laws then we could write them in the form (see e.g. |2|)

$$-S_N \in \partial j(u_N) \tag{6}$$

where ∂ denotes the subdifferentiation and j is the convex superpotential of the structure-support interaction, which is lower semicontinuous (l.s.c.) and proper function on R (for definitions we refer to [2] or to [6]). This law generalizes the classical potential law. Indeed if j is differentiable, (6) may be put in the form

$$- S_N = dj\,(u_N)\,/\,du_N \tag{6a}$$

The notion of superpotential, which has been introduced by J.J. Moreau in 1968 [7], has been generalized |2|,|4| for the case of nonconvex energy functions by means of the generalized gradient of Clarke |8|, (see |2| or Appendix in |4|). Thus we can easily verify that the nonmonotone laws of fig. 2 can be written in the form

$$-S_N \in \bar{\partial}\phi\,(u_N) \tag{7}$$

where ϕ is a function with values in $[-\infty, +\infty]$, and $\bar{\partial}$ is the symbol of generalized gradient. Note that if ϕ is convex, l.s.c. and proper, then $\bar{\partial}\phi = \partial\phi$. Thus the law of fig. 2a may be put in the form (7) where

$$\phi\,(u_N) = \{\int_0^{u_N} k(\xi)\,d\xi = \bar{\Phi}\,(u_N) \text{ if } u_N > 0, \text{ and } 0 \text{ if } u_N \leq 0\} = \int_0^{u_N^+} K(\xi)\,d\xi \tag{8}$$

Here u_N^+ is the positive part of u_N, i.e. $u_N^+ = \sup\,(0, u_N)$.
Analogously, for the law of fig. 2b.

$$\phi\,(u_N) = \{\int_\alpha^{u_N} k(\xi)d\xi \text{ if } u_N \geq \alpha, \text{ and } 0 \text{ if } u_N < \alpha\} \tag{9}$$

$\phi\,(u_N)$ is numerically the area formed between the curve $k(u_N)$ and the u_N - axis. By the definition of $\bar{\partial}$,(7) is equivalent to the hemivariational inequality

$$\phi^{\uparrow}(u_N, v_N - u_N) \geq -S_N\,(v_N - u_N) \quad \forall\, v_N \in \mathbb{R} \tag{10}$$

which we will use later. Here $\phi^{\uparrow}(.,.)$ denotes the upper-subderivative [8], which is a generalization of the classical differential, has been recently introduced in nonsmooth analysis, and is given in the present case by the formula

$$\phi^{\uparrow}(u_N, v_N) = \limsup\,[\phi\,(u_N + h + \lambda v_N) - \phi(u_N + h)]/\lambda \quad \text{as } \lambda \to 0_+,\ h \to 0. \tag{11}$$

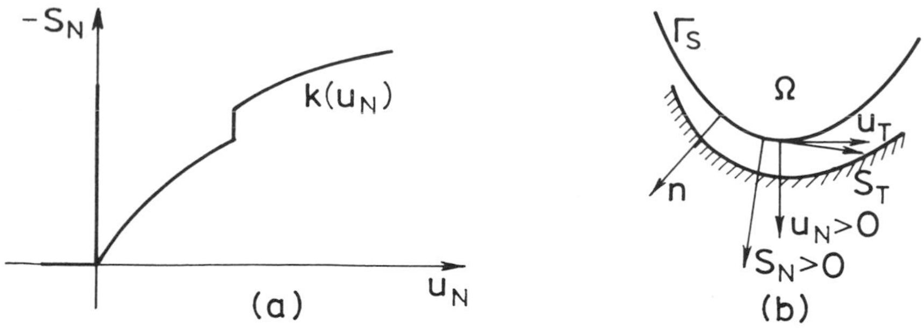

Fig. 1. Monotone Unilateral Contact Laws

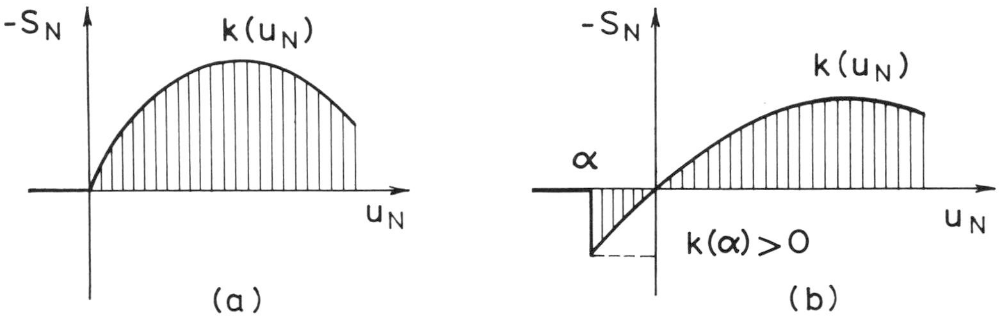

Fig. 2. Nonmonotone Unilateral and Adhesive Contact Laws

Monotone and Nonmonotone Frictional Contact Laws

In the tangential direction we may combine the aforementioned laws with classical (bilateral) laws, e.g. (3). Of course we may also consider monotone relations of the form

$$-S_T \in \partial \tilde{j} \ (u_T) \tag{12}$$

or, more general nonmonotone relations of the form

$$-S_T \in \bar{\partial} \ \tilde{\phi} \ (u_T) \tag{13}$$

in order to describe more complicated phenomena in the tangential direction, as for instance, frictional phenomena. It is well-known [1], that the Coulomb friction boundary condition has on Γ_S the form:

$$\text{if } |S_T| < \mu \ |S_N|, \text{ then } u_{Ti} = 0 \tag{14a}$$

if $|S_T| = \mu\ |S_N|$, then there exists $\lambda \geq 0$ such that

$$u_{Ti} = -\lambda\ S_{Ti}\ , \quad i = 1, 2, 3 \tag{14b}$$

Here $\mu>0$ is the coefficient of friction. If we assume for the present that

$$S_N = C_N, \quad C_N = C_N(x) \quad \text{on } \Gamma_S \tag{15}$$

then, (14a,b) can be put in the form (12), where
$\tilde{j}(u_T)=\mu|C_N||u_T|$ and $\partial j(.)$ has the graph of Fig. 3a (for Ω twodimensional). We may analogously consider the orthotropic friction [9]. If α and β are the orthotropy directions on Γ_S, μ_α, μ_β the corresponding coefficients of friction and (15) holds, then we may consider the superpotential

$$\tilde{j}(u_T) = |C_N|(\mu_\alpha^2\ u_{T\alpha}^2 + \mu_\beta^2\ u_{T\beta}^2)^{\frac{1}{2}} \tag{16}$$

and (14a,b) are replaced by the relation

$$\text{If} \quad [(\frac{S_{T\alpha}}{\mu_\alpha})^2 + (\frac{S_{T\beta}}{\mu_\beta})^2]^{\frac{1}{2}} < |C_N|, \quad \text{then } u_{T\alpha} = u_{T\beta} = 0 \tag{17a}$$

$$\text{if} \quad [(\frac{S_{T\alpha}}{\mu_\alpha})^2 + (\frac{S_{T\beta}}{\mu_\beta})^2]^{\frac{1}{2}} = |C_N|, \quad \text{then there exists } \lambda\geq 0 \text{ such that}$$

$$u_{Ti} = -\lambda\frac{S_{Ti}}{\mu_i^2} \quad i = \alpha,\beta \tag{17b}$$

More generally we may introduce the anisotropic friction [2] by considering on Γ_S a general limit condition of the form $f(S_T;S_N)<0$ where f is a convex continuously differentiable function of S_{Ti}, i=1,2. Then we write the friction law in the form (cf [2] p. 189)

$$-u_T \in \partial I_K\ (S_T;S_N) \tag{18}$$

where $I_K(S_{Ti}\ ;\ S_N) = \{0 \text{ if } S_T \in K,\ \infty \text{ otherwise}\}$ is the indicator of the convex closed set $K = \{S_T|f(S_T\ ;\ S_N) \leq 0\}$. The subdifferential is formed with respect to S_T. Then (18) can be written in the form

$$u_{Ti} = -\lambda\ \frac{\partial f(S_T\ ;\ S_N)}{\partial S_{Ti}} \quad i = 1,\ 2 \quad \lambda\geq 0,\ f\leq 0,\ \lambda f=0, \tag{19}$$

where S_N is a parameter of the problem. Obviously the inverse form of (18) is

$$S_T \in \partial I_K^C\ (-u_T\ ;\ S_N)\ , \tag{20}$$

where I_K^c denotes the conjugate functional. If

$$f(S_T;S_N) = [(\frac{S_{T\alpha}}{\mu_\alpha})^2 + (\frac{S_{T\beta}}{\mu_\beta})^2]^{\frac{1}{2}} - |C_N| \tag{21}$$

we obtain again (17a,b).

If S_N is not given, and the contact is unilateral, then obviously frictional effects will appear only on the contact regions. Therefore, we may write the nonmonotone unilateral contact boundary condition with monotone friction in the form,

$$-S_N \in \bar{\partial}\phi(u_N) \quad \text{and} \quad \text{if } S_N = 0, \quad \text{then } S_T = 0, \tag{22a}$$

$$\text{if } S_N < 0, \quad \text{then } - S_T \in \partial j\ (u_T;S_N) \tag{22b}$$

For instance, in the simpler case of (4) we have:

$$\text{if} \quad u_N < 0, \quad \text{then} \quad S_N = 0 \quad \text{and } S_T = 0 \tag{23a}$$

$$\text{if} \quad u_N \geq 0, \quad \text{then } -S_N \in \bar{\partial}\bar{\phi}(u_N) \quad \text{and } -S_T \in \partial j\ (u_T;S_N) \tag{23b}$$

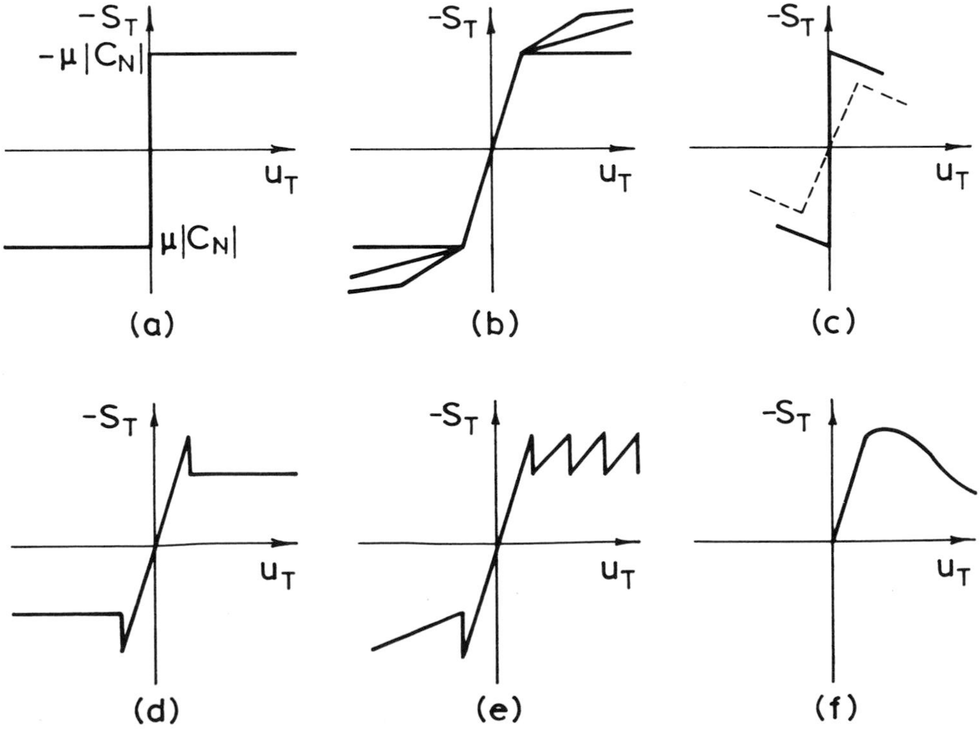

Fig. 3. Friction Boundary Conditions

Besides the aforementioned laws we must consider generalizations having as generic diagrams those of figs 3b. These laws can also be put in the form (12) with S_N as parameter.

Let us assume again that (15) holds. Then certain more realistic laws of friction are depicted (for Ω twodimensional) in Figs. 3c,d,e. We mention also fig. 3f describing the friction forces between reinforcement and concrete. These laws are nonmonotone and therefore they can be put in the form (13) where $\tilde{\varphi}$ is the area between the graph of the $(-S_T, u_T)$ law and the u_T-axis. It is obvious that the $(-S_T, u_T)$ graph may include vertical jumps describing local cracking and crushing of the asperities on the contact surface. Note here the conceptual analogy between Scalon's diagram (tension stiffening effect for the reinforced concrete in tension [10][11]) and the sawteeth diagram of fig. 3e describing the local change of mechanical properties on the contact surface. Also worth noting is the more general analogy between the strain-stress diagram for granular materials (including worksoftening) and the $(-S_T, u_T)$ diagram.

Nonmonotone friction laws can be obtained by considering a continuously differentiable function $\tilde{f}=\tilde{f}(S_T;S_N)$ such that $f(S_T;S_N)< 0$ is the friction criterion (which reminds us of a holonomic plasticity yield function [2]). Let $\tilde{K}$ be more generally the closed nonconvex subset of the S_T-space and $I_{\tilde{K}}(S_T;S_N)$ be the indicator of $\tilde{K}$. Thus we write the friction law in the form

$$-u_T \in \bar{\partial} I_{\tilde{K}}(S_T;S_N) = N_{\tilde{K}}(S_T;S_N), \tag{24}$$

where $N_{\tilde{K}}$ denotes the normal cone to $\tilde{K}$ in the sense of Clarke [8] (Fig. 4a). If $T_{\tilde{K}}(S_T;S_N)$ denotes the corresponding tangential cone, then (24) is equivalent to the hemivariational inequality (cf. [2] p. 158).

$$I_{T_{\tilde{K}}(S_T;S_N)}(S_T^* - S_T) \geq -u_{Ti}(S_{Ti}^* - S_{Ti}) \;\forall S_T^* \in \mathbb{R}^3. \tag{25}$$

Here I donotes again the indicator. Accordingly, if for a value of S_N,

$$(S_T^* - S_T) \in T_{\tilde{K}}(-S_T;S_N) \tag{26}$$

then (25) becomes

$$-u_{Ti}(S_{Ti}^* - S_{Ti}) \leq 0 \;\forall S_T^* \in \tilde{K} \tag{27}$$

recalling Hill's principle in elastoplasticity. Inclusion (24) is also equivalent to the assertion that for every value of S_N, $u_T S_T$ is substationary [2] with respect to $\tilde{K}$ at the point S_T. Note that (26) holds if $\tilde{K}$ does not have reentrant corners as is the case if $K = \{S_T | \tilde{f}(S_T;S_N) \leq 0\}$ (Fig. 4b). Then if $0 \notin \bar{\partial} f(S_T;S_N)$, (24) is equivalent to the classical linear complementarity expressions

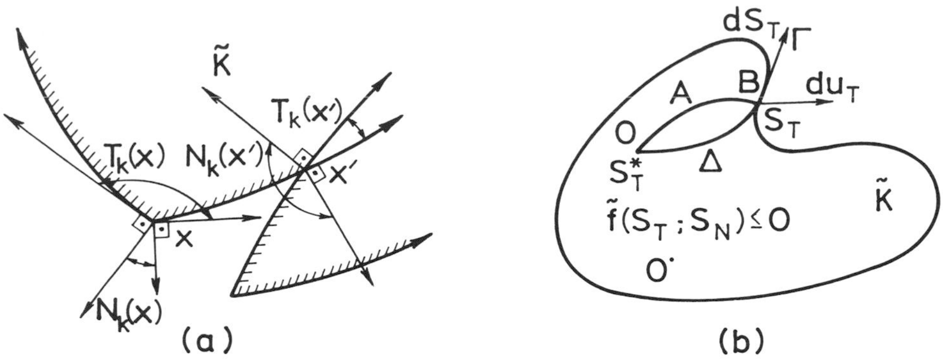

Fig. 4. Normal cones and the Nonconvexity in Friction Laws.

$$-u_{Ti} = \frac{\lambda\, \partial \tilde{f}\,(S_T;S_N)}{\partial\, S_{Ti}}\,,\quad \lambda \geq 0,\quad \tilde{f} \leq 0,\quad \lambda\tilde{f}=0\,. \tag{28}$$

If, more generally, the sliding-adhesive friction boundary is the boundary of the set $\tilde{K}$ defined by the inequalities $\tilde{f}_n\,(S_T;S_N) \leq 0$ $n = 1, 2, \ldots m$ and each f_n is continuously differentiable then (24) is again equivalent-on the assumtion that $0 \notin \bar{\partial}\,(\max_n \tilde{f}_n\,(S_T;S_N))$ - to the expression

$$-u_{Ti} = \sum_{n=1}^{m} \lambda_n \frac{\partial \tilde{f}_n(S_T;S_N)}{\partial\, S_{Ti}}\,,\quad \lambda_n \geq 0,\quad \tilde{f}_n \leq 0,\ \lambda_n\, \tilde{f}_n = 0 \tag{28a}$$

The analogy between friction and plasticity is well-known. Then (24) corresponds to nonmonotone rigid-plastic effects. Following the ideas of [2] ch. 4 we can write nonmonotone friction laws corresponding to elastic rigid plastic holonomic laws. Generally speaking we may consider the relation (13) where $\tilde{\varphi}$ depends on $S_N=S_N(u)$ and we may write

$$-S_T \in \bar{\partial}\tilde{\phi}\,(u_T;S_N)\,. \tag{29}$$

With respect to (29) we can adapt the idea of the non-local effects induced by the asperities of the contact surface [12] . Then we have to replace S_N by $Q(S_N(u))$, where Q is an appropriate mollifier (smoothing operator). Law (29) may describe the effect of the local crushing and cracking of the asperities (three dimensional generalization of fig 3e-analogy to Scanlon's diagram in concrete). This can be obtained optimally by superimposing, e.g. to a classical orthotropic friction law, a nonconvex superpotential law implying a local sawteeth behaviour (cf. e.g. (4.3.75) of [2] p. 155 and its obvious generalization in two dimensions). This is a general purpose law which of course needs certain additional information on the contact surface, like the magnitude of the local jump of the stresses due to the cracking of the asperities.

The aforementioned static friction laws are useful for purely static problems and certain classes of quasi-static problems. Note that these laws can be parallelized to the law of the deformation theory of plasticity. Of course the existing comparison and criticism between this theory and the flow theory of plasticity applies also to some extent to the static friction laws. For dynamic problems we may consider friction laws introducing a relation between S_T and v_T where v_T denotes the tangential velocity [1] . With respect to this general and more rational-from the standpoint of mechanics- framework we can write all the aforementioned laws simply by replacing u_T by v_T. In this context we can explain the nonconvexity of the sliding-adhesive friction boundary by means of exactly analogous thoughts to those of Green and Naghdi in plasticity [13] and we can assume a star-shaped (nonconvex) friction boundary. Moreover, a nonmonotone (S_T, v_T) law of the form

$$-v_T \in \partial w\,(S_T\ ;\ S_N) \tag{30}$$

is justified on the general assumption that the properties of the asperities change during the tangential displacement. The proof is analogous to the one given for the nonconvex plasticity in [2] p. 154: let us consider a $\overset{*}{S}_T$ on or inside the friction surface (Fig. 4b) and an external agency changing the stress vector and let S_T be its current value such that $\tilde{f}\,(S_T;S_N)=0$. The external agency produces further a tangential displacement increment $-du_T$ and let dS_T be the corresponding reaction increment tangential to the boundary of $\tilde{K}$.$0 \in \mathrm{int}\,R$ (i.e the zero tangential force does not produce sliding). Further the loading releases dS_T and the traction returns to its initial value $\overset{*}{S}_T$. We assume that for the loading path OABΔO, for which no sliding occurs, the work done consists of two parts: one conservative (i.e. it does not depend on the path) and one nonconvervative. The first may correspond to the elastic behaviour of the asperities causing reversible deformation, whereas the second to the possible change of the elastic properties due to the local fracture effects. We denote the nonrecoverable work by $\tilde{w}$ and we assume that it is for a given S_N a function of S_T, $\overset{*}{S}_T$, and the path followed. Moreover $\tilde{w}=0$ for $\overset{*}{S}_T=S_T$. Accordingly the work produced over the loading path OABΓΔO, assumed to be positive, is

$$-(S_{Ti}-S^*_{Ti})\,du_{Ti}+dS_{Ti}du_{Ti}+\tilde{w}(S^*_T,S_T,\ \text{path};S_N) > 0. \tag{31}$$

(31) holds for every $S^*_T \in \tilde{K}$ and for every path which does not pass out of $\tilde{K}$. Setting $S_T=S^*_T$ in (31) implies $dS_{Ti}du_{Ti} > 0$, and since this expression can be made as small as we wish, we obtain the inequality

$$- (S_{Ti}-S^{\star}_{Ti})\ du_{Ti} + \tilde{w}\ (S^{\star}_{T}, S_{T}, \text{path}; S_{N}) \geq 0 \quad \forall S^{\star}_{T} \in \tilde{K} \tag{32}$$

Now we assume that a function w: $R^3 \to [-\infty, +\infty]$ exists such that

$$w\uparrow (S_T, S^{\star}_T - S_T ; S_N) = \{ \tilde{w}\ (S_T , S^{\star}_T, \text{path}; S_N) \text{ if } S_T , S^{\star}_T \in \tilde{K} \text{ and} \quad \infty \quad \text{otherwise} \} \tag{33}$$

where $w^{\uparrow}(.,.;S_N)$ denotes the upper-subderivative (cf. [8] and [2] p. 143) of w. By the definition of ∂w we obtain from (32) and(33) that

$$-du_T \in \bar{\partial} w\ (S_T ; S_N). \tag{34}$$

In place of the increment du_T we may introduce the velocity v_T and (30) results. Finally we would like to note that the nonmonotone theory of friction can be directly connected to the generalized (nonconvex) hypothesis of normal dissipation (cf. [2] p. 157) after an appropriate definition of the hidden variables. In this context see also [14].

BOUNDARY VALUE PROBLEMS AND HEMIVARIATIONAL INEQUALITIES

The Nonmonotone Frictional Contact Problems

In this section we assume for the present that on Γ_S S_N is given, i.e. (15) holds, whereas the frictional forces are connected to the displacements by (29). Moreover on Γ_U and on Γ_F (1) and (2) hold. On the assumption of small strains we can write in Ω the equations of equilibrium and the strain-displacement equations

$$\sigma_{ij,j} + f_i = 0 \quad \text{in } \Omega \quad , \tag{35}$$

$$\varepsilon_{ij}\ (u) = \frac{1}{2} (u_{i,j} + u_{j,i}) \quad \text{in } \Omega \tag{36}$$

where $f=\{f_i\}$ denotes the volume force vector, $\sigma = \{\sigma_{ij}\}$ (reps. $\varepsilon = \{\varepsilon_{ij}\}$ is the stress (resp. the strain) tensor and the comma denotes the partial derivative. Moreover, let

$$\sigma_{ij} = C_{ijhk}\ \varepsilon_{hk} \quad \text{in } \Omega , \tag{37}$$

where $C=\{C_{ijhk}\}$ i, j, h, k = 1,2,3 is the Hooke tensor of elasticity with the well-known symmetry and ellipticity properties. We denote also by a (.,.) the bilinear form of the elasticity

$$\alpha\ (u, v) = \int C_{ijhk}\ \varepsilon_{ij}(u)\ \varepsilon_{hk}\ (v)\ d\Omega \tag{38}$$

From (35) (36) and (1), (2) we derive by means of the Green-Gauss theorem the relation

$$\int_\Omega \sigma_{ij}(\varepsilon_{ij}(v) - \varepsilon_{ij}(u)\, d\Omega = \int_\Omega f_i(v_i - u_i)\, d\Omega + \int_{\Gamma_F} F_i(v_i - u_i)\, d\Gamma +$$

$$\int_{\Gamma_S} [S_{Ti}(v_{Ti} - u_{Ti}) + S_N (v_N - u_N)]\, d\Gamma \quad \forall v \in X \tag{39}$$

where X is the kinematically admissible set

$$X = \{v | v_i = U_i \text{ on } \Gamma_U \quad \text{and} \quad \tilde{\phi}\,(v_T\ ;\ C_N) < \infty \quad \text{on} \quad \Gamma_S\} \tag{40}$$

and v denotes in this section a perturbed value of the displacement u. (29) is equivalent to the hemivariational inequality

$$\tilde{\phi}^\uparrow\,(u_T,\ v_T - u_T\ ;\ S_N) \geq -\ S_{Ti}\,(v_{Ti} - u_{Ti}) \quad \forall v_T \in \mathbb{R}^3 \tag{41}$$

which, when combined with (39) and (37) (38) and (15) leads to the following problem : Find $u \in X$ such that

$$\alpha(u,\ v-u) + \int_{\Gamma_S} \tilde{\phi}^\uparrow(u_T,\ v_T - u_T;\ C_N)d\Gamma \geq \Omega \int_\Omega f_i(v_i - u_i)d\Omega + \int_{\Gamma_S} C_N\ (v_N - u_N)d\Gamma +$$

$$+ \int_{\Gamma_F} F_i\ (v_i - u_i)\ d\Gamma \quad \forall v \in X \tag{42}$$

This hemivariational inequality includes all the monotone friction laws (for $S_N = C_N$) as special cases.

Note that if S_N is not given then in (42) $\tilde{\phi}^\uparrow$ depends on the unknown function $S_N(u)$. For instance, if we have contact with an undeformable support (Signorini-Fichera problem with nonmonotone friction), then we introduce the kinematically admissible set

$$X_1 = \{v | v_i = U_i \text{ on } \Gamma_U,\ v_N \leq 0 \text{ and } \tilde{\phi}\,(v_T; S_N(v)) < \infty \text{ on } \Gamma_S\} \tag{43}$$

and we have to find $u \in X_1$ such that

$$\alpha(u, v-u) + \int_{\Gamma_S} \tilde{\phi}^\uparrow(u_T,\ v_T - u_T; S_N(u))\ d\Gamma \geq \int_\Omega f_i(v_i - u_i)d\Omega + \int_{\Gamma_F} F_i(v_i - u_i)d\Gamma \ \forall v \in X_1 \tag{44}$$

We assume further that (29) holds but with $\tilde{\phi}$ independent of S_N. If moreover, in the normal direction to Γ_S a monotone $S_N - u_N$ law holds, which due to (6) is equivalent to the variational inequality,

$$j\,(v_N) - j(u_N) \geq -S_N(v_N - u_N) \quad \forall v_N \in \mathbb{R} \tag{45}$$

we are led to the following variational-hemivariational inequality: Find $u \in X_2 = \{v | v \in X,\ j\,(v_N) < \infty \text{ on } \Gamma_S\}$ such that

$$\alpha(u, v-u) + \int_{\Gamma_S} [\tilde{\phi}^{\uparrow}(u_T, v_T-u_T) + (v_N)-j(u_N)]d\Gamma \geq \int_{\Omega} f_i(v_i-u_i)d\Omega +$$

$$+ \int_{\Gamma_F} F_i (v_i-u_i) d\Gamma \; \forall v \in X_2 \tag{46}$$

If (37) is replaced by monotone or nonmonotone holonomic stress-strain relations

$$\sigma \in \partial W (\varepsilon) \tag{47}$$

$$\sigma \in \partial W (\varepsilon) \tag{48}$$

respectively [2] [4] , then (42), (44) and (46) still hold with the following modifications : a) If (47) holds then $\alpha(u, v-u)$ is replaced by $\int_{\Omega} [W(\varepsilon(v)) - W(\varepsilon(u))]d\Omega$ and in all the kinematically admissible sets the inequality $W(\varepsilon) < \infty$ has to be included. So, e.g. in the case of locking materials this last inequality is equivalent to $Q(\varepsilon) \leq 0$, where $Q=Q(\varepsilon)$ is the locking criterion. b) If (48) holds, then $\alpha(u, v-u)$ is replaced by $\int_{\Omega} W^{\uparrow} (\varepsilon(u), \varepsilon(v)-\varepsilon(u))d\Omega$ and the inequality $W(\varepsilon)<\infty$ is included in the sets X, X_1, X_2. (47) represents the deformation theory of plasticity, Hencky materials, elastic perfectly plastic (holonomic) materials, elastic ideally locking materials, materials with polygonal stress-strain law e.t.c. (cf. [2] [15]); whereas (48) describes the behaviour of materials with nonmonotone stress-strain behaviour, like concrete, soils, granular media, composite materials,the Scanlon's effect in reinforced concrete e.t.c. (cf. [2] , [4],[10]).

The Nonmonotone Unilateral Contact Problem with Friction

In this section we assume for the present that S_T is given on Γ_S, i.e. that (3) holds together with the nonconvex unilateral contact boundary condition (7). If (1), (2), (35), (36) and (37), hold we are led to the following problem: Find $u \in \tilde{X}$ such that

$$\alpha (u, v-u) + \int_{\Gamma_S} \phi^{\uparrow}(u_N, v_N-u_N)d\Gamma \geq \int_{\Omega} f_i(v_i-u_i)d\Omega + \int_{\Gamma_S} C_{Ti}(v_{Ti}-u_{Ti})d\Gamma +$$

$$+ \int_{\Gamma_F} F_i (v_i-u_i) d\Gamma \quad \forall v \in \tilde{X} \tag{49}$$

where $\tilde{X}=\{v|v_i = U_i \text{ on } \Gamma_U , \phi(v_N) < \infty \text{ on } \Gamma_S\}$

If in the tangential direction instead of (3) the relation (12) (resp. (13)) holds, then we are led to the problem: find $u \in \tilde{X}_1$ such that

$$\alpha(u, v-u)+ \int_{\Gamma_S} [\phi^{\uparrow}(u_N, v_N-u_N)+(\tilde{j}(v_T)-\tilde{j}(u_T)) \text{ (resp. } \tilde{\phi}^{\uparrow}(u_T, v_T-u_T)]d\Gamma$$

$$\geq \int_{\Omega} f_i(v_i-u_i) d\Omega + \int_{\Gamma_F} F_i(v_i-u_i)d\Gamma \quad \forall v \in \tilde{X}_1 \tag{50}$$

where $\tilde{X}_1 = \{v \mid v \in \tilde{X} , \tilde{j}(v_T) < \infty \quad (\text{resp } \tilde{\phi}(v_T) < \infty)\}$

If instead of (37), (47) or (48) hold the hemivariational inequalities (49) and (50) are modified exactly as in the previous section. Until now we have not considered the coupled and more realistic boundary condition (22) or its nonmonotone generalization:

If $u_N < 0$, then $S_N=0$ and $S_T= 0$ (51a)

if $u_N \geq 0$, then $-S_N \in \bar{\partial}\bar{\phi}(u_N)$ and $-S_T \in \bar{\partial}\tilde{\phi}(u_T;S_N)$ (51b)

Such coupled laws present, even in the monotone case, considerable difficulties both from the standpoint of their numerical and their mathematical treatment. Thus, the proof for the existence and the approximation of the solution of (44) with $\tilde{\phi}(u_T)=j(u_T)=\mu|S_N(u)\|u_T|$ (i.e. the classical monotone Coulomb's friction) has recently been achieved for the case of small friction by the use of a fixed point method, which actually was the mathematical version of a heuristic procedure proposed in [17][6] in a more general context. The same method is proposed here for the nonmonotone coupled boundary conditions (51) describing the unilateral frictional contact: In the first step we assume that S_N is given, say $S_N^{(1)}$ and for the value $C_N=S_N^{(1)}$ we solve the corresponding nonmonotone friction problem, i.e. the hemivariational inequality (42). Let the calculated value of S_T be $S_T^{(1)}$. For this value the nonmonotone unilateral contact problem is solved, i.e. the hemivariational inequality (49) for $C_T=S_T^{(1)}$, and a value of S_N, say $S_N^{(2)}$ is obtained which is transmitted into (42) and so on, until the differences $S_N^{(\alpha)}-S_N^{(\alpha-1)}$ and $S_T^{(\alpha)}-S_T^{(\alpha-1)}$ become sufficiently small. This algorithm decomposes the coupled problem into two "known" problems and has all the features of a fixed point algorithm.

On the Solution of the Hemivariational Inequalities. Applications

Both the hemivariational inequalities (42) and (49) are of the same nature and for them some more rigorous mathematical results can be proved concerning the existence and the approximation of the solution. Here we deal with the case $\Omega \subset \mathbb{R}^2$ and Γ Lipschitzian. For $\Omega \subset \mathbb{R}^3$ (49) does not present any additional difficulty, in contrast to (42), where the method of proof needs a thorough modification. For the sake of simplicity we assume that on Γ_U $u_i=0$ (otherwise an appropriate translation is necessary). Moreover, let $f_i \in L^2(\Omega)$, C_N, $C_{Ti} \in L^2(\Gamma_S)$, $F_i \in L^2(\Gamma_F)$, $C_{ijhk} \in L^\infty(\Omega)$ and u_i, $v_i \in V = \{v_i \mid v_i \in H^1(\Omega),\ v_i = 0 \text{ on } \Gamma_U\}$. We assume that the multivalued functions $\bar{\partial}\phi$ and $\bar{\partial}\tilde{\phi}$ are generated by the generally discontinuous functions b , $\tilde{b} \in L^\infty_{loc}(R)$ by "filling in" their discontinuities: for a $\delta > 0$ and $\xi \in R$ we define:

$$\bar{b}_{\delta}(\xi) = \underset{|\xi-\xi_1|<\delta}{\text{essup}}\; b\,(\xi_1) \quad \text{and} \quad \underline{b}_{\delta}(\xi) = \underset{|\xi-\xi_1|<\delta}{\text{essinf}}\; b\,(\xi_1) \tag{52}$$

and then we take the limits, as $\delta\to 0$. These limits $\bar{b}\,(\xi) = \lim_{\delta\to 0} \bar{b}_{\delta}\,(\xi)$ and $\underline{b}\,(\xi) = \lim_{\delta\to 0} \underline{b}_{\delta}(\xi)$, exist due to the monotonicity properties of $\bar{b}_{\delta}(\xi)$ and $\underline{b}_{\delta}(\xi)$ when considered as functions of δ. Then let us consider the multivalued function

$$\hat{b}\,(\xi) = [\underline{b}(\xi),\, \bar{b}\,(\xi)] \tag{53}$$

where $[.,.]$ denotes a closed interval. It can be shown [18][2], that, if $b(\xi_{\pm 0})$ exists at every ξ, then a locally Lipschitz continuous function ϕ can be determined up to a constant by the formula $\phi(\xi) = \int_0^{\xi} b\,(\xi_1)\,d\xi_1$, such that

$$\hat{b}\,(\xi) = \bar{\partial}\,\phi\,(\xi)\,. \tag{54}$$

Then (11) supplies the upper subderivative $\phi^{\uparrow}(\xi,z)$. Analogously, from $\tilde{b}$ we obtain $\tilde{\phi}$. We shall further define the regularized problems (42ε) and (49ε) corresponding to (42) and (49). We introduce the nollifier p, i.e.

$p \in C_0^{\infty}\,(-1, +1)$ $p > 0$ and $\int_{-\infty}^{+\infty} p(\xi)d\xi = 1$ (see e.g. |19|) and let

$$p_{\varepsilon}\,(\xi) = \frac{1}{\varepsilon}\,p\,(\frac{\xi}{\varepsilon}) \qquad \varepsilon > 0\,. \tag{55}$$

Then we form the convolutions

$$b_{\varepsilon} = p \star b \quad \text{and} \quad \tilde{b}_{\varepsilon} = p \star \tilde{b} \tag{56}$$

and we define the regularized problems (42_{ε}) and (49_{ε}): Find $u_{\varepsilon_i} \in V$ such that

$$\alpha\,(u_{\varepsilon}, v) + \int_{\Gamma_S} \tilde{b}_{\varepsilon}(u_{T\varepsilon})\, v_T\, d\Gamma = \int_{\Omega} f_i v_i d\Omega + \int_{\Gamma_S} C_N v_i d\Gamma + \int_{\Gamma_F} F_i v_i d\Gamma \;\; \forall v_i \in V \tag{42ε}$$

and

$$\alpha\,(u_{\varepsilon}, v) + \int_{\Gamma_S} b_{\varepsilon}(u_{N\varepsilon})\, v_N d\Gamma = \int_{\Omega} f_i v_i d\Omega + \int_{\Gamma_S} C_{Ti} v_{Ti} d\Gamma + \int_{\Gamma_F} F_i v_i d\Gamma \;\forall v_i \in V \tag{49ε}$$

In order to discretize (42ε) and (49ε) we consider a Galerkin basis of V and let V_n be the corresponding n-dimensional subspace. The corresponding finite dimensional problems ($42\varepsilon n$) and ($49\varepsilon n$) result simply from ($42\varepsilon n$) by replacing u_{ε} by $u_{\varepsilon n}$ and V by V_n. The following proposition holds:

Proposition: Suppose that for some ξ

$$\underset{(-\infty,-\xi)}{\text{essup}}\; b\,(\xi) \leq \underset{(\xi,\, +\infty)}{\text{ess inf}}\; b\,(\xi) \tag{57}$$

and similarly for $\tilde{b}$. Then ($42\varepsilon n$) and ($49\varepsilon n$) each have a solution $u_{\varepsilon n}$, which, as $\varepsilon\to 0$ and $n\to\infty$, tends weakly in VxVxV to the solution u of (42) and (49) respectively.

The proof which uses compactness arguments follows the same steps as the proof of prop. 8.4.1 in [2] and is omitted here. Moreover, we can prove that, if there is a constant $c>0$ and $q>1$ such that

$$|b(\xi)| \le c\,(1+|\xi|^q) \quad \forall\, \xi \in \mathbb{R} \tag{58}$$

and similarly for $\tilde{b}$, then $u_{\varepsilon n}$ tends strongly to u in VxVxV. All the above conditions are satisfied in most practical cases (e.g. for Fig. 3). The finite dimensional form of the hemivariational inequalities can be realized by an appropriate finite element scheme. Moreover, a more heuristic regularization can be applied. The regularized finite dimensional problems ($42\varepsilon n$) and ($49\varepsilon n$) are nonlinear algebraic equations with sparse nonlinear terms and their solution can be obtained by an appropriate algorithm, for instance, of the Newton-Raphson type. It should be noted that the regularization can be avoided if the chosen algorithm can accomodate the nonlinearities arising from the "filled in" jumps. In this case the discretized model is considered with additional fictive springs having as stress-strain relations the $(-S_N, u_N)$ or $(-S_T, u_T)$ diagrams. It should be noted that the solution is generally nonunique due to the lack of monotonicity. There is also another procedure inherently similar to the aforementioned, which consists of the minimization of the potential energy corresponding to (42) and (49) over the kinematically admissible set. As shown in [4] [2] the potential energy, e.g. for (42),

$$\Pi(u) = \frac{1}{2}\,\alpha(u,u) + \int_{\Gamma_S} \tilde{\phi}(u_T)\,d\Gamma - \int_{\Omega} f_i u_i\, d\Omega - \int_{\Gamma_F} F_i u_i\, d\Gamma - \int_{\Gamma_S} C_N(u_N)\, d\Gamma \tag{59}$$

is substationary at the position of equilibrium with respect to X and moreover, every local minimum of Π over X is a substationarity point. In order to find this local minimum a nonconvex optimization algorithm is necessary. Finally, we would like to point out that the hemivariational inequalities express from the standpoint of physics the principle of virtual work.

In the case of dynamic friction laws, variations of the velocities are considered leading to similar types of hemivariational inequalities as in the static case. Obviously, if the displacements are small, f_i is replaced by $f_i - \rho\frac{\partial^2 u_i}{\partial t^2}$ where ρ denotes the density of the body and t the time variable.

As a first application we consider the nonconvex contact problem of the symmetric beam of Fig. 5. The displacements are obtained by a modified Newton-Raphson method applied directly to the discretized structure. In Fig. 5b the regularized symmetric reaction-displacement diagram is depicted. The calculation is repeated for several values of ε, and the solution is obtained as $\varepsilon\to 0$ (actual-

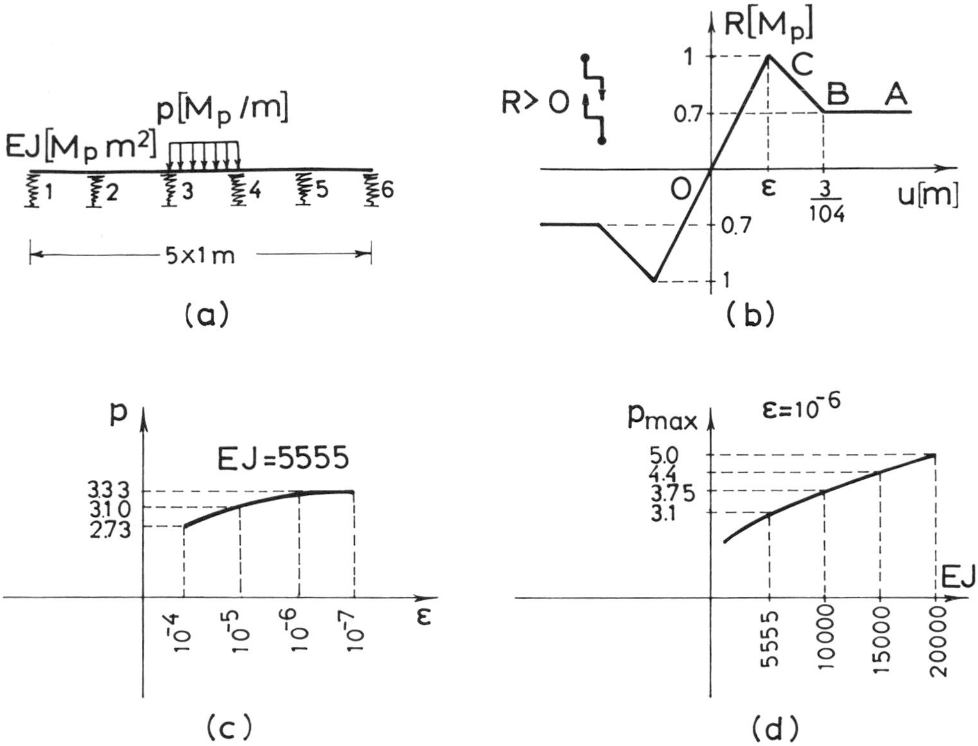

Fig. 5. A Nonconvex Contact Problem

ly for $\varepsilon=10^{-7}$). For this value of p (and for all the others in Fig. 5c) all the springs are on the branch AB of the diagram. For the same load, say p=3.0, we note that the springs 1,2,3 are for $\varepsilon=10^{-7}$ on the branches OC, OC, CB respectively-as well as for $\varepsilon=10^{-6}$, $\varepsilon=10^{-5}$, whereas for $\varepsilon=10^{-4}$ all the springs are on the branch BA. For the smaller loading p=2.80 and for $\varepsilon=10^{-4}$ the first spring is on OC, whereas 2 and 3 on BA.

As a second example we consider the frictional contact problem of Fig. 6. Part I is linear elastic and part II rigid. On the contact surface we assume that the friction diagram of Fig. 6b holds and that $u_N=0$. This condition is introduced into the kinematically admissible set. The hemivariational inequality is regularized and as $\varepsilon\to 0$ (actually for $\varepsilon=10^{-6}$) the solution is obtained. Finally, it should be noted that most of the questions concerning the rigorous numerical analysis of hemivariational inequalities are still open.

Acknowledgements

The author would like to thank Dr. H. O. May and Mr. E. Koltsakis for helpful discussions and for the numerical implementation.

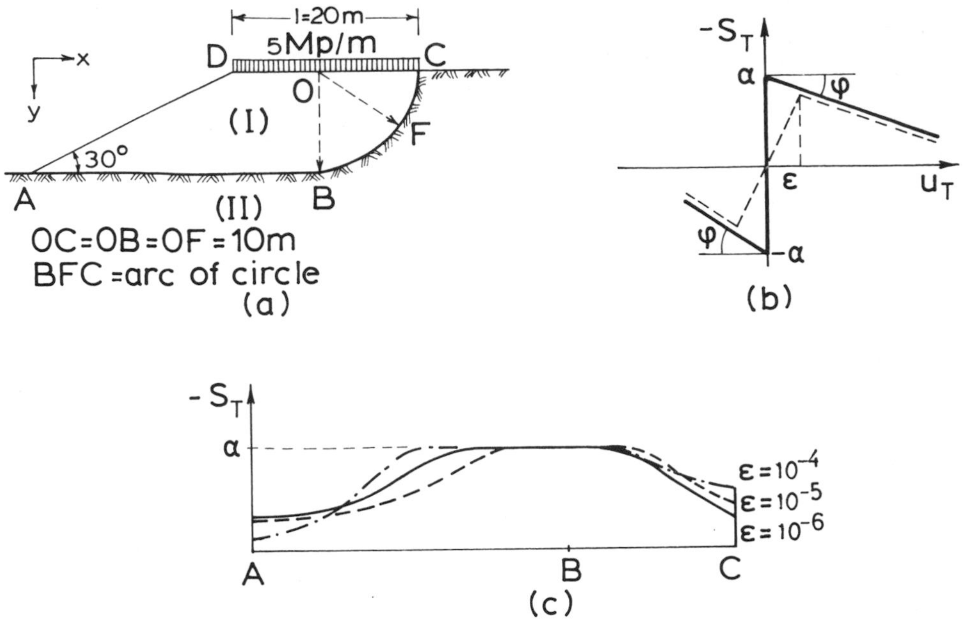

Fig. 6. A Nonmonotone Friction Problem ($E=1\ Mp/cm^2$, $\nu=0.35$, $\alpha=1 Mp/m^2$, $\varphi=15^0$)

References

1 G. Duvaut and J.L. Lions, Les inéquations en Mécanique et en Physique Dunod, Paris 1972.

2 P.D. Panagiotopoulos, Inequality Problems in Mechanics and Applications. Convex and Nonconvex Energy Functions, Birkhäuser Verlag, Boston 1985.

3 G. Fichera, Problem elastotatici con vincoli unilaterali: il problema di Signorini con ambigue condizioni al contorno. Mem. Accad. Naz. Lincei, VII 7, 1964, 91-140.

4 P.D. Panagiotopoulos, Non-Convex Energy Functions. Hemivariational Inequalities and Substationarity Principles. Acta Mechanica 42 (1983) 160-183.

5 P.D. Panagiotopoulos and A. Avdelas, A Hemivariational Inequality Approach to the Unilateral Contact Problem and Substationarity Principles, Ing. Archiv 54 (1984) 401-412.

6 P.D. Panagiotopoulos, A Variational Inequality Approach to the Friction Problem of Structures with Convex Strain Energy Density and Application to the Frictional Unilateral Contact Problem. J. Struct. Mech. 6 (1978) 303-318.

7 J.J. Moreau, La notion de sur-potentiel et les liaisons unilatérales en elastostatique. C.R. Acad. Sc. Paris 267A (1968) 954-957.

8 R.T. Rockafellar, Generalized Directional Derivatives and Subgradients of Non-convex Functions. Can. J. Math. XXXII (1980) 257-280.

9 R. Michalowski and Z. Mróz, Associated and Non-associated Sliding Rules in Contact Friction Problems. Arch. of Mech. (Arch. Mech. Stosowanej) 30 (1978) 259-276.

10 P.D. Panagiotopoulos and C. Baniotopoulos, A Hemivariational Inequality and Substationarity Approach to the Interface Problem: Theory and Prospects of Applications. Eng. Anal. 1 (1984) 20-31.

11 H. Floegl and H.A. Mang, Tension Stiffening Concept Based on Bond Slip. ASCE, ST 12, 108 (1982) 2681-2701.

12 E.B. Pires and J.T. Oden, Analysis of Contact Problems with Friction Under Oscillating Loads, Comp. Meth. Appl. Mech. Eng. 39 (1983) 337-362.

13 A.E. Green and P.M. Naghdi, A General Theory of an Elastic-Plastic Continuum. Arch. Rat. Mech. Anal. 18 (1965) 251-281.

14 A. Klarbring, General Contact Boundary Conditions and the Analysis of Frictional Systems, To appear.

15 F. Lené, Sur les matériaux élastiques á énergie de déformation non quadratique. J. de Mécanique 13 (1974) 499-534.

16 J. Nečas, J. Jarušek and J. Haslinger, On the Solution of the Variational Inequality to the Signorini Problem with Small Friction. Bulletino U.M.I. 17B (1980) 796-811.

17 P.D. Panagiotopoulos, A Nonlinear Programming Approach to the Unilateral Contact-and Friction- Boundary Value Problem in the Theory of Elasticity. Ing. Archiv 44 (1975) 421-432.

18 K.C. Chang, Variational Methods for Non-differentiable Functionals and their Applications to Partial Differential Equations. J. Math. Anal. Appl. 80 (1981) 102-129.

19 J.T. Oden and J.N. Reddy, An Introduction to the Mathematical Theory of Finite Elements, J. Wiley, N. York 1976.

THE INFLUENCE OF SLIP HARDENING AND INTERFACE COMPLIANCE ON CONTACT STRESS DISTRIBUTIONS. A MATHEMATICAL PROGRAMMING APPROACH

A. KLARBRING

Linköping Institute of Technology, Department of Mechanical Engineering, S-581 83 Linköping (Sweden)

ABSTRACT

The present study deals with elastic contact problems with friction. A new model of frictional interface behaviour is proposed, which is analogous to that previously used to describe elasto-plastic non-associated piecewise linear material behaviour. Some works which experimentally support the model are also reviewed.

In order to justify a finite dimensional approximation of the interface model, a variational formulation in combination with numerical integration is utilized. When this finite dimensional approximation is combined with a matrix equation, describing the elastic behaviour of contacting bodies, one obtains what has been termed a parametric Linear Complementarity Problem (LCP) involving derivatives. The author has previously modified a solution algorithm for parametric LCPs involving derivatives to apply to frictional problems and in the last section of the paper this algorithm is used to analyse a punch indentation problem. The influence of slip hardening and interface compliance on contact stress distributions are investigated.

1. INTRODUCTION

Frictional effects are of prime importance in a number of engineering applications. Investigations of dry surface friction have, consequently, often been the subject of experimental research. However, when it comes to numerical and analytical work on solid mechanics problems that involve surface friction, nearly all investigations have been limited to the use of the friction law of Coulomb, devised in the 18th century. Indeed, this must be considered as a natural consequence of the absence of generally accepted phenomenological models of friction. However, recently several investigators have put forward modifications of Coulomb's friction law: Oden and Pires [1] introduce nonlocal and nonlinear friction laws and Curnier [2] proposes an extension of the concept of "standard generalized materials" that constitutes a law of "standard generalized friction". The latter work is an extension of the works by Fredriksson [3] and by Michalowski and Mroz [4]. Moreover, it is shown in Ref. [5] that convex and non-convex superpotentials can be used to describe and generalize Curnier's law of standard generalized friction.

A motivation behind the recent work on phenomenological friction laws has been the fact that experiments show a similarity between frictional force-displacement relations and elasto-plastic stress-strain curves. The present work also utilizes this observation; and, furthermore, it makes use of the fact that mathematical programming methods have been found useful for the description of elasto-plastic stress-strain relations and the performance of numerical analyses of elasto-plastic structures [6]. Thus, relations similar to those governing piecewise linear non-associated plasticity will be assumed to describe the interface behaviour of contact surfaces. The approximations and assumptions involved in such models are discussed in Section 2, where the problem to be treated is introduced. In Section 3 a variational formulation is given, which enables us to give a finite dimensional approximation of the interface model. When combined with the relations of linear elasticity this finite dimensional description results in a mathematical programming problem known as a parametric Linear Complementarity Problem (LCP) involving derivatives. A solution method for this problem has been suggested in Ref. [7]. Finally, in Section 4, using the proposed method, some features of the model introduced are numerically studied. A problem of obvious theoretical and practical interest is treated. It concerns the loading and unloading of a flat circular punch contacting a half-space, and it has previously been treated by Spence [8], Turner [9] and Torstenfelt [10], assuming Coulomb friction.

2. THE ELASTIC CONTACT PROBLEM WITH A GENERAL FRICTION LAW

Consider two linear elastic bodies A and B, contacting each other over a part of their boundaries. Let $\underset{\sim}{u}^{A(B)}$ denote the displacement and $\underset{\sim}{\nu}^{A(B)}$ the outward boundary normal vector of body A(B). Using cartesian vector components and the summation convention for indices i and j, for each point of a contact surface S_c we define relative normal and tangential contact displacements as

$$w_N = - [\nu_i^A u_i^A + \nu_i^B u_i^B], \tag{2.1}$$

$$\underset{\sim}{w}_T = \underset{\sim}{u}_T^A - \underset{\sim}{u}_T^B, \tag{2.2}$$

where $u_{Ti}^{A(B)} = u_i^{A(B)} - \nu_i^{A(B)} \nu_j^{A(B)} u_j^{A(B)}$. Furthermore, define normal and tangential contact stresses as

$$p_N = \sigma_{ij}^A \nu_i^A \nu_j^A = \sigma_{ij}^B \nu_i^B \nu_j^B, \tag{2.3}$$

$$\underset{\sim}{p}_T = \underset{\sim}{\sigma}_T^B = - \underset{\sim}{\sigma}_T^A, \tag{2.4}$$

where $\sigma_{Ti}^{A(B)} = \sigma_{ij}^{A(B)} \nu_j^{A(B)} - p_N \nu_i^{A(B)}$ and where σ_{ij} are the components of the stress tensor.

For every point on S_c the following set of relations are assumed to be valid

$$p_N = Q_N w_N, \tag{2.5}$$

$$\underset{\sim}{w}_T = \underset{\sim}{w}_T^a + \underset{\sim}{w}_T^s, \tag{2.6}$$

$$\underset{\sim}{p}_T = \underset{\sim}{Q}_T \underset{\sim}{w}_T^a, \tag{2.7}$$

$$\dot{w}_{Ti}^s = \sum_{\alpha=1}^{n} \dot{\lambda}_\alpha \frac{\partial \phi_\alpha}{\partial p_{Ti}}, \tag{2.8}$$

$$\phi_\alpha(p_N, \underset{\sim}{p}_T, \underset{\sim}{A}) \leq 0, \ \alpha = 1, \ldots, n, \tag{2.9}$$

$$\dot{\lambda}_\alpha \geq 0, \ \dot{\lambda}_\alpha \phi_\alpha(p_N, \underset{\sim}{p}_T, \underset{\sim}{A}) = 0, \ \alpha = 1, \ldots, n, \tag{2.10}$$

where $(\dot{\ })$ denotes time derivative. These relations describe a constitutive contact boundary behavior, featuring reversible and non-reversible tangential displacements $\underset{\sim}{w}_T^a$ and $\underset{\sim}{w}_T^s$, and normal and tangential contact surface layer stiffnesses Q_N and $\underset{\sim}{Q}_T$. Furthermore, the non-reversible part of the tangential displacement rate $\dot{\underset{\sim}{w}}_T^s$ is governed by slip functions ϕ_α and non-negative multiplier rates $\dot{\lambda}_\alpha$, which should satisfy the complementarity conditions (2.9) and (2.10). $\underset{\sim}{A}$ is a vector of internal force variables which indicates the previous contact stress history.

Equation (2.5) describes the normal stiffness behavior of the contact interface. Experimental research and proposed theoretical models of such behavior, for rough surfaces of metallic bodies, have been reviewed by Back et al [11] and more recently by Oden and Martins [12]. These summaries show that the relation between normal contact stress and relative normal contact displacement is nonlinear. Contact stresses below 5 MPa are shown to be proportional to a power, in the range 2 to 3.33, of the relative normal displacement and a logaritmic relation holds for larger loads. These observations show that the linear relation (2.5) must be regarded as an approximation of the real behavior. However, for small variations of normal contact stress and for qualitative analysis it should be useful. Moreover, Taniguchi et al [13] recently proposed and experimentally verified a linear relation for contact stresses above 10 MPa. The situation described is illustrated in Fig. 2.1.

Equation (2.6) divides the relative tangential displacement into two parts, one reversible and one non-reversible. An alternative terminology would be the adhesion and slip parts of the tangential displacement [2]. This division is suggested by several experimental investigations [14] - [16] and it is also predicted by micromechanical models of the contact surface [17], see Fig. 2.2. Equation (2.7) describes a linear relation between the tangential contact stress

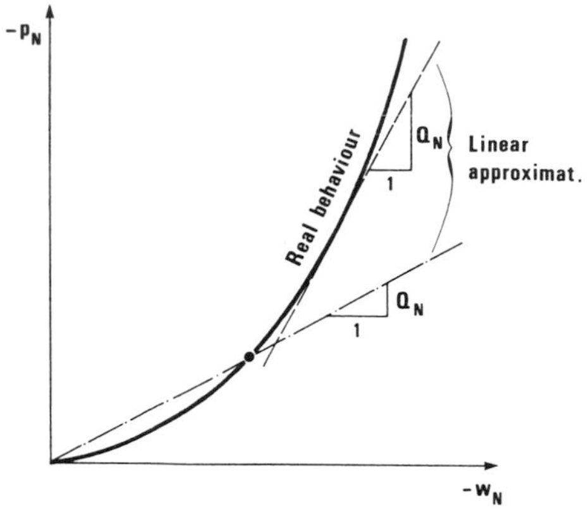

Fig. 2.1. Normal load-deflection characteristic of interface.

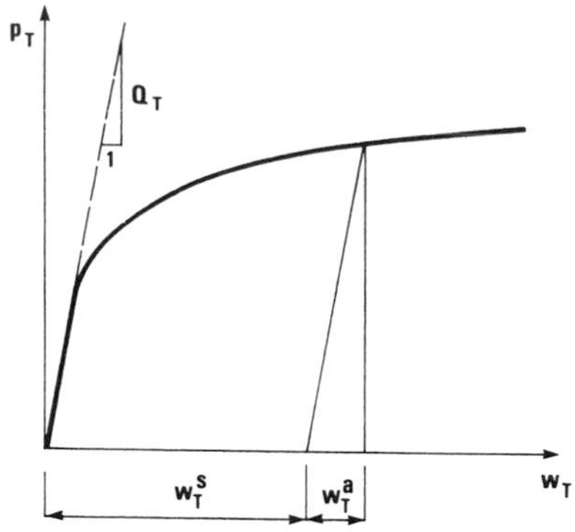

Fig. 2.2. Tangential load-deflection characteristic of interface.

and the reversible part of the tangential displacement. This is a realistic assumption provided that the normal contact stress is approximately constant. The qualitative relation between the shear stiffness $\underset{\sim}{Q}_T$ and the normal contact stress is shown in Fig. 2.3. For machined surfaces of metallic bodies, the ratio between the shear stiffness and the normal stiffness at corresponding normal stress is shown by experimental data to be about 0.8. Furthermore, the shear stiffness is shown by Lindgren [16] to have a definite influence on the global stress and displacement states of some types of mechanical joints and it should therefore be incorporated in a model of mechanical interface behavior.

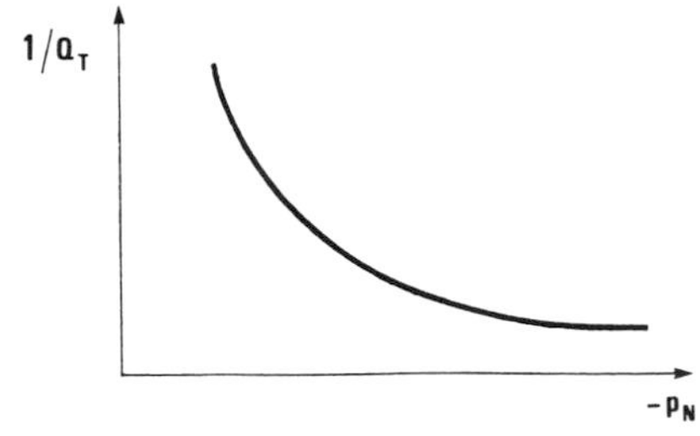

Fig. 2.3. Relation between shear stiffness and normal contact stress.

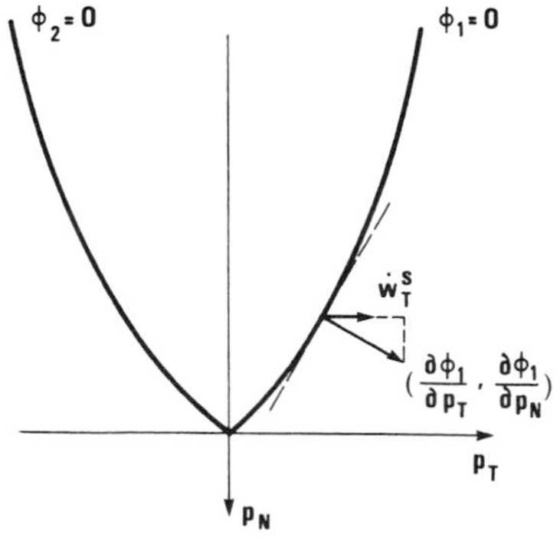

Fig. 2.4. Slip functions in two dimensions.

The non-reversible part of the relative tangential displacement is governed by relations (2.8) through (2.10). No slip takes place at a given point of the contact surface if the contact stress state at that point is such that the slip function ϕ_α is less than zero for every $\alpha = 1, \ldots, n$. If, for one index α, $\phi_\alpha \leq 0$ is satisfied as an equality, the non-reversible slip rate increment will be directed along the projection of the normal of the surface $\phi_\alpha = 0$ on the plane p_N = constant, as shown if Fig. 2.4 for a two-dimensional situation. The magnitude of the slip rate increment is determined by the non-negative multiplier rate $\dot{\lambda}_\alpha$. If several slip functions equal zero simultaneously, the slip rate increment will be a linear combination of the respective projections of normal vectors as described by equation (2.8).

The slip functions at a point of the contact surface will in general depend on the previous slip history at that point. In (2.9) this is indicated by the presence of the internal forces $\underset{\sim}{A}$ as variables of ϕ_α. It is shown in Ref. [5] that Curnier's law of standard generalized friction gives rise to a linear relation between $\dot{\underset{\sim}{A}}$ and the time rate of change of the n-vector $\underset{\sim}{\lambda}^t = [\lambda_1, \ldots, \lambda_n]$ of non-negative multipliers. This indicates that the following relation holds:

$$\dot{\phi}_\alpha = \frac{\partial \phi_\alpha}{\partial p_N} \dot{p}_N + \frac{\partial \phi_\alpha}{\partial p_{Ti}} \dot{p}_{Ti} - \underset{\sim}{H}_\alpha \dot{\underset{\sim}{\lambda}}, \tag{2.11}$$

where $\underset{\sim}{H}_\alpha$ is a vector which describes the hardening behavior of the slip function ϕ_α due to the rate of slip at the particular point of the contact surface. In this paper we will assume that ϕ_α is linearly dependent on p_N, $\underset{\sim}{p}_T$ and $\underset{\sim}{\lambda}$; therefore, (2.11) can be integrated to yield

$$\phi_\alpha = \frac{\partial \phi_\alpha}{\partial p_N} p_N + \frac{\partial \phi_\alpha}{\partial p_{Ti}} p_{Ti} - K_\alpha - \underset{\sim}{H}_\alpha \underset{\sim}{\lambda}, \tag{2.12}$$

where K_α are constants which, when $\underset{\sim}{\lambda} = \underset{\sim}{0}$, represent the ortogonal distance between the plane $\phi_\alpha = 0$ and the "origin" $p_N = 0$, $\underset{\sim}{p}_T = \underset{\sim}{0}$ in contact stress space. Hardening has previously been incorporated in friction calculations by Fredriksson [3] and Lindgren [16].

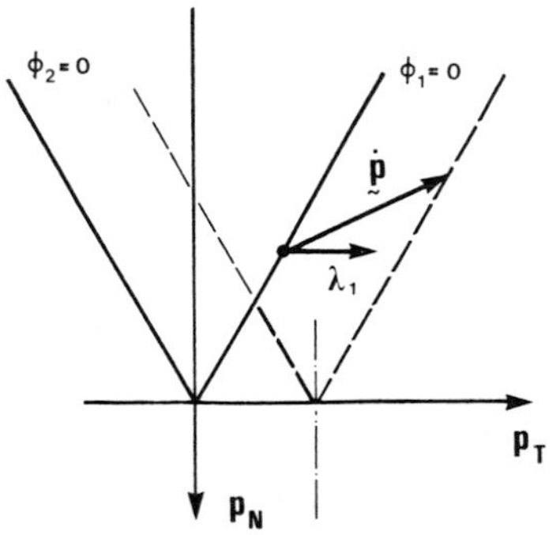

Fig. 2.5. Coulomb's law with hardening in two dimensions.

Example: In a two-dimensional situation, a generalisation of Coulomb's law to involve hardening can be given by the following slip functions:

$$\phi_1 = p_T + \mu p_N - H\lambda_1 + H\lambda_2, \tag{2.13}$$

$$\phi_2 = - p_T + \mu p_N + H\lambda_1 - H\lambda_2, \tag{2.14}$$

where μ is the coefficient of friction and H is a hardening modulus. An illustration of this situation is given in Fig. 2.5.

Finally, to complete the description of the problem, apart from interface conditions, classical relations of linear elasticity are assumed to be valid. That is, the bodies could be subjected to body forces over their volume, and surface forces and prescribed displacements over the part of the boundaries not in contact. Furthermore, since this is an evolution problem proper initial displacement conditions must be prescribed.

3. VARIATIONAL FORMULATION, DISCRETIZATION AND SOLUTION METHOD

Variational formulations of frictionless contact problems have during the last decades occurred in a great number of publications. Variational formulations where friction is incorporated in a physically admissible way have, on the other hand, been scarce [1]. The mathematical questions connected with such formulations have only recently been investigated [18]. In this paper we will make use of variational arguments to an extent necessary for the justification of a finite dimensional approximation of the friction law.

To that end, introduce a space P of contact stress fields over S_c and its dual W, the space of relative contact displacement fields over S_c. Also, the n-vector fields of multipliers belong to a space Λ. Note that vector notation will be used to denote both physical vectors and fields of that vector. No misinterpretation should be possible. The non-negative multipliers belong to the set

$$\Lambda^+ = \{\underset{\sim}{\gamma} \in \Lambda | \gamma_\alpha \geq 0 \text{ on } S_c, \ \alpha = 1, \ldots, n\}.$$

Furthermore, assume that P is decomposed into disjoint spaces P_N and P_T, such that P_N consists of fields of normal contact stresses and P_T of tangential contact stresses. The dual space W is similarly decomposed.

Let the initial conditions be such that $\underset{\sim}{w}_T^S$ and $\underset{\sim}{\lambda}$ can be assumed to be zero. Together with the linearity assumption on all ϕ_α this implies that (2.8) may be integrated, simply by removing the dots. If Q_N and $\underset{\sim}{Q}_T$ are invertible, the friction law as described by (2.5) through (2.10) and (2.12) can then be represented by the following variational statement: for $\underset{\sim}{p} \in P$, $\underset{\sim}{w} \in W$ and $\underset{\sim}{\lambda} \in \Lambda^+$ it holds that

$$\int_{S_c} (Q_N^{-1} p_N - w_N) s_N dS = 0, \quad \forall\, s_N \in P_N, \tag{3.1}$$

$$\int_{S_c} (Q_{Tij}^{-1} p_{Tj} + \sum_{\alpha=1}^{n} \lambda_\alpha \frac{\partial\phi_\alpha}{\partial p_{Ti}} - w_{Ti}) s_{Ti} dS = 0, \quad \forall\, \underset{\sim}{s}_T \in P_T, \tag{3.2}$$

$$\int_{S_c} (\frac{\partial\phi_\alpha}{\partial p_N} p_N + \frac{\partial\phi_\alpha}{\partial p_{Ti}} p_{Ti} - K_\alpha - \sum_{\beta=1}^{n} H_{\alpha\beta}\lambda_\beta)(\dot{\lambda}_\alpha - \gamma_\alpha) dS \geq 0, \ \alpha = 1, \ldots, n,$$

$$\forall\, \underset{\sim}{\gamma} \in \Lambda^+. \tag{3.3}$$

To show that the complementarity conditions (2.9) and (2.10) are equivalent to the variational inequality (3.3) is a straightforward exercise.

For computational purposes the spaces used above should be replaced by finite dimensional approximations. To that end the integrals will be evaluated using numerical integration. Let f(x) be a function defined on S_c. The integral of f is then given by

$$I[f] = \sum_{e=1}^{E} I_e[f], \ I_e[f] = \sum_{g=1}^{G} W_g^e f(x_g^e). \tag{3.4}$$

Here E is the number of contact boundary elements, G is the number of integration points, x_g^e, of the quadrature rule used and W_g^e are the weighting factors.

As has been pointed out in a similar context by Oden and Kikuchi [19], finite dimensional approximations of P, W and Λ can be associated with the quadrature rule used for numerical integration. Denote these spaces by subscript h. If, for example, $\underset{\sim}{s}^h$ belongs to P_h, the generalized coordinates of such an approximation may be taken to be the values $s_i^h(x_g^e)$ of the field at the integration points. Furthermore, for the variational inequality (3.3) to make sense in a finite dimensional setting, an approximation of the set Λ^+ should be given. This is obtained by imposing the inequality conditions on $\lambda^h \in \Lambda_h$ at the integration points, i e

$$\Lambda_h^+ = \{\underset{\sim}{\gamma}^h \in \Lambda_h | \gamma_\alpha^h(x_g^e) \geq 0, \ \alpha = 1, \ldots, n, \ 1 \leq e \leq E, \ 1 \leq g \leq G\}.$$

Numerical integration of the statements (3.1) through (3.3) yields

$$\sum_{e=1}^{E} \sum_{g=1}^{G} W_g^e (Q_N^{-1} p_N^h(x_g^e) - w_N^h(x_g^e)) s_N^h(x_g^e) = \sum_{e=1}^{E} \sum_{g=1}^{G} (\frac{Q_N^{-1}}{W_g^e} R_{Ng}^e - r_{Ng}^e) R_{Ng}^{*e} = 0,$$

$$\forall\, \underset{\sim}{s}_N^h \in P_{Nh}, \tag{3.5}$$

$$\sum_{e=1}^{E} \sum_{g=1}^{G} W_g^e (Q_{Tij}^{-1} p_{Tj}^h(x_g^e) + \sum_{\alpha=1}^{n} \lambda_\alpha^h(x_g^e) \frac{\partial \phi_\alpha}{\partial p_{Ti}} - w_{Ti}^h(x_g^e)) s_{Ti}^h(x_g^e) =$$

$$= \sum_{e=1}^{E} \sum_{g=1}^{G} (\frac{Q_{Tij}^{-1}}{W_g^e} R_{Tjg}^e + \sum_{\alpha=1}^{n} \lambda_\alpha^h(x_g^e) \frac{\partial \phi_\alpha}{\partial p_{Ti}} - r_{Tig}^e) R_{Tig}^{*e} = 0, \quad \forall s_T^h \in P_{Th}, \qquad (3.6)$$

$$\sum_{e=1}^{E} \sum_{g=1}^{G} W_g^e (\frac{\partial \phi_\alpha}{\partial p_N} p_N^h(x_g^e) + \frac{\partial \phi_\alpha}{\partial p_{Ti}} p_{Ti}^h(x_g^e) - K_\alpha(x_g^e) - \sum_{\beta=1}^{n} H_{\alpha\beta} \lambda_\beta^h(x_g^e)) \cdot$$

$$\cdot (\dot{\lambda}_\alpha^h(x_g^e) - \gamma_\alpha^h(x_g^e)) = \sum_{e=1}^{E} \sum_{g=1}^{G} (\frac{\partial \phi_\alpha}{\partial p_N} R_{Ng}^e + \frac{\partial \phi_\alpha}{\partial p_{Ti}} R_{Tig}^e - W_g^e K_\alpha(x_g^e) -$$

$$- W_g^e \sum_{\beta=1}^{n} H_{\alpha\beta} \lambda_\beta^h(x_g^e))(\dot{\lambda}_\alpha^h(x_g^e) - \gamma_\alpha^h(x_g^e)) \geq 0, \quad \alpha = 1, \dots, n, \quad \forall \gamma^h \in \Lambda_h^+, \qquad (3.7)$$

where we have introduced the notations

$$R_{Ng}^e = W_g^e p_N^h(x_g^e), \quad R_{Ng}^{*e} = W_g^e s_n^h(x_g^e), \quad R_{Tig}^e = W_g^e p_{Ti}^h(x_g^e), \quad R_{Tig}^{*e} = W_g^e s_{Ti}^h(x_g^e),$$

$$r_{Ng}^e = w_N^h(x_g^e), \quad r_{Tig}^e = w_{Ti}^h(x_g^e).$$

Note that R_{Ng}^e, etc, represents contact forces and r_{Ng}^e, etc, represents the associated contact displacements.

It is now evident that the approximation of the friction law can be given a marix form. Introduce column vectors $\underset{\sim}{R}$, $\underset{\sim}{R}_N$ and $\underset{\sim}{R}_T$ of contact force and of normal and tangential contact forces; and, similarly, column vectors $\underset{\sim}{r}$, $\underset{\sim}{r}_N$ and $\underset{\sim}{r}_T$ of the associated contact displacement variables. Assume that they are referred to local coordinate systems so that the compositions $\underset{\sim}{R}^t = [\underset{\sim}{R}_N^t, \underset{\sim}{R}_T^t]$ and $\underset{\sim}{r}^t = [\underset{\sim}{r}_N^t, \underset{\sim}{r}_T^t]$ hold. Furthermore, introduce block diagonal matrices

$$\bar{Q}_N^{-1} = \text{diag} \{\frac{Q_N^{-1}}{W_1^1}, \dots, \frac{Q_N^{-1}}{W_g^e}, \dots, \frac{Q_N^{-1}}{W_G^E}\},$$

$$\bar{Q}_T^{-1} = \text{diag} \{[\frac{Q_T^{-1}}{W_1^1}], \dots, [\frac{Q_T^{-1}}{W_g^e}], \dots, [\frac{Q_T^{-1}}{W_G^E}]\},$$

$$\underset{\sim}{N}_N = \text{diag} \{[\frac{\partial \phi_1}{\partial p_N}, \dots, \frac{\partial \phi_n}{\partial p_N}]_1^1, \dots, [\frac{\partial \phi_1}{\partial p_N}, \dots]_g^e, \dots, [\frac{\partial \phi_1}{\partial p_N}, \dots]_G^E\},$$

$$\underset{\sim}{N}_T = \text{diag} \{[\nabla \phi_1, \dots, \nabla \phi_n]_1^1, \dots, [\nabla \phi_1, \dots]_g^e, \dots, [\nabla \phi_1, \dots]_G^E\},$$

$$\bar{\underset{\sim}{H}} = \text{diag} \{W_1^1 \underset{\sim}{H}, \dots, W_g^e \underset{\sim}{H}, \dots, W_G^E \underset{\sim}{H}\}.$$

Here, $[\ ]_g^e$ stands for evaluation at integration point x_g^e and $\nabla\phi_i$ is the gradient of ϕ_i with respect to $\underset{\sim}{p}_T$, referred to the local coordinate systems. From the above matrices, construct the block matrices $\bar{\underset{\sim}{Q}}^{-1} = \text{diag}\{\bar{\underset{\sim}{Q}}_N^{-1}, \bar{\underset{\sim}{Q}}_T^{-1}\}$, $\underset{\sim}{N}^t = [\underset{\sim}{N}_N^t, \underset{\sim}{N}_T^t]$ and $\overset{o}{\underset{\sim}{N}}{}^t = [\underset{\sim}{0}, \underset{\sim}{N}_T^t]$, and assume from now on that K_α equals zero as is the case in the generalisation of Coulomb's friction law given in the Example of Section 2.

A vector-matrix form of (3.5) through (3.7) is now given by

$$\bar{\underset{\sim}{Q}}^{-1}\underset{\sim}{R} + \overset{o}{\underset{\sim}{N}}\underset{\sim}{\lambda} = \underset{\sim}{r}, \tag{3.8}$$

$$\underset{\sim}{\phi} = \underset{\sim}{N}^t\underset{\sim}{R} - \bar{\underset{\sim}{H}}\underset{\sim}{\lambda} \leq \underset{\sim}{0}, \tag{3.9}$$

$$\underset{\sim}{\phi}^t\dot{\underset{\sim}{\lambda}} = 0,\ \dot{\underset{\sim}{\lambda}} \geq \underset{\sim}{0}, \tag{3.10}$$

where we have used vectors $\underset{\sim}{\phi}$ and $\underset{\sim}{\lambda}$ of dimension $n \cdot E \cdot G$.

In Ref. [20] three-dimensional contact problems with friction but without slip hardening and interface compliance have been dealt with. It was there shown that a finite element discretization of the linear elastic bodies involved may be employed to obtain a relation between $\underset{\sim}{R}$, $\underset{\sim}{r}$ and the prescribed forces and displacements on the structure. Under reasonable assumptions [20] on the support of the bodies and on the compatibility between the discretizations of bodies and interface the following relation was shown

$$\underset{\sim}{R} = \underset{\sim}{P} - \underset{\sim}{\kappa}\underset{\sim}{r}, \tag{3.11}$$

where $\underset{\sim}{\kappa}$ is a positive semi-definite matrix and $\underset{\sim}{P}$ represents the actions on the structure.

Introducing now the notation $\underset{\sim}{r}^s = \overset{o}{\underset{\sim}{N}}\underset{\sim}{\lambda}$ for the non-reversible part of $\underset{\sim}{r}$, relation (3.8) results in the stiffness relation

$$\begin{bmatrix} \bar{\underset{\sim}{Q}}, & -\bar{\underset{\sim}{Q}} \\ -\bar{\underset{\sim}{Q}}, & \bar{\underset{\sim}{Q}} \end{bmatrix} \begin{bmatrix} \underset{\sim}{r} \\ \underset{\sim}{r}^s \end{bmatrix} = \begin{bmatrix} \underset{\sim}{R} \\ -\underset{\sim}{R} \end{bmatrix} \tag{3.12}$$

This relation can be combined with (3.11) and $\underset{\sim}{r}$ can be eliminated by using a substructure method. We obtain

$$\underset{\sim}{R} = \bar{\underset{\sim}{P}} - \bar{\underset{\sim}{\kappa}}\underset{\sim}{r}^s, \tag{3.13}$$

where $\bar{\underset{\sim}{\kappa}} = \bar{\underset{\sim}{Q}} - \bar{\underset{\sim}{Q}}(\underset{\sim}{\kappa} + \bar{\underset{\sim}{Q}})^{-1}\bar{\underset{\sim}{Q}}$ and $\bar{\underset{\sim}{P}} = \bar{\underset{\sim}{Q}}(\underset{\sim}{\kappa} + \bar{\underset{\sim}{Q}})^{-1}\underset{\sim}{P}$. Combining (3.13) with (3.9) and (3.10) we obtain

$$\underset{\sim}{\phi} = \underset{\sim}{N}^t\underset{\sim}{\bar{P}} - (\underset{\sim}{N}^t\underset{\sim}{\bar{\kappa}}\underset{\sim}{N} + \underset{\sim}{\bar{H}})\underset{\sim}{\lambda} \leq \underset{\sim}{0}, \tag{3.14}$$

$$\underset{\sim}{\phi}^t\underset{\sim}{\dot{\lambda}} = 0, \; \underset{\sim}{\dot{\lambda}} \geq \underset{\sim}{0}. \tag{3.15}$$

This is a Linear Complementary Problem (LCP) involving derivatives. A parametric version of this problem was mathematically studied by Kaneko [7] who proposed an algorithm for its solution. In Ref. [20] the author modified Kaneko's analysis to apply to the case of contact problems with friction. Conditions which assure the existence of a unique solution of our problem were given. In the next section Kaneko's method will be used for the solution of a punch indentation problem.

4. APPLICATION

As an application of the theory presented in the previous sections we will consider the indentation of a half-space by a flat circular-cylindrical punch, Fig. 4.1. Both bodies are assumed to be isotropic, homogeneous and linear elastic with different elastic constants. The punch is given a prescribed displacement as indicated in Fig. 4.1. We will treat both the problem of a displacement that increases monotonically to a final value δ and the problem of gradually removing this displacement. The unloading is monitored by a multiplier β, taking values between one and zero.

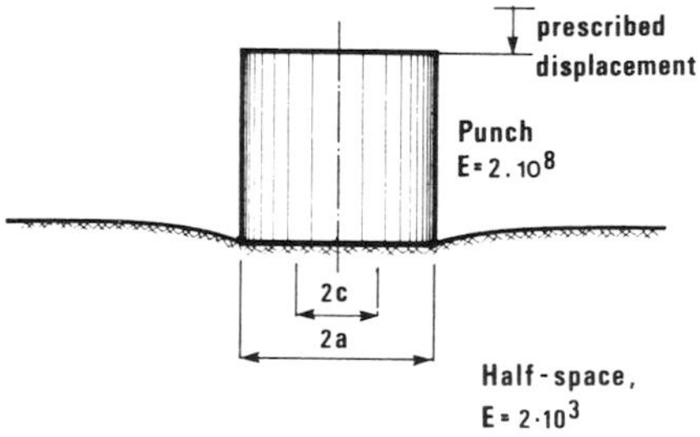

Fig. 4.1. Problem treated.

In the case where the punch is rigid and the contact is frictionless, this problem was solved by Boussinesq in 1885. If Coulomb friction is assumed a more complicated problem has to be treated. However, for a monotonically increasing load a closed form solution has been obtained by Spence [8] under the assumption that the normal contact pressure is unaffected by the presence of friction. Using an iterative numerical method he also treated the problem without this restriction. These results show that the loading problem is linear and that, surrounding an adhesive circle $r \leq c$, there develops a region $c < r \leq a$ where inward slip takes place.

On the other hand, the unloading problem remained unsolved by the classical approach used by Spence. The reason was that it was not possible to specify in advance the way in which the contact surface was to be divided into regions of adhesion and slip. In Ref. [21] Spence shows that the assumption, suggested by intuition, that reverse slip begins at the outher edge of the contact region does not produce a solution. A variational formulation of the problem overcomes this difficulty, since for such a formulation the actual distribution of regions of adhesion and slip will be a product of the computation. This fact was used by Turner [9] who showed that a minimization of a complementary energy functional yields the solution of the problem if one assumes the same simplifications as enabled Spence to obtain a closed form solution in the loading case. Turner's numerical results show that two regions of slip separated by a region of adhesion develop during the unloading phase. Moreover, recently Torstenfelt [10] has considered both the loading and unloading problems using an iterative finite element technique. His calculations show that the simplifying assumptions used by Turner do not effect the results in any crucial way.

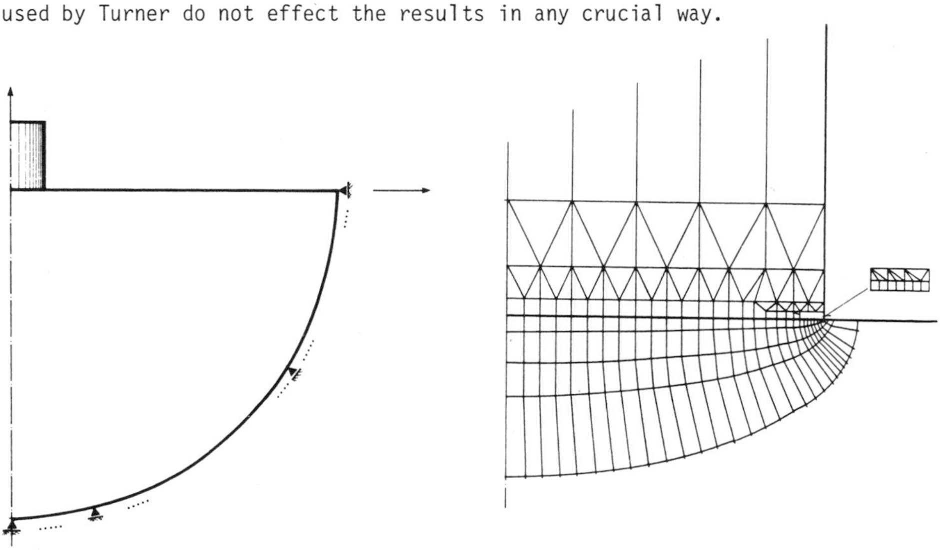

Fig. 4.2. Finite element model. Finite "half-space" and mesh in region of contact.

In this paper, for the problem described above, we will examine the influence of slip hardening and interface compliance on the contact stress distribution and load-penetration relation. Departure from the assumption of punch rigidity will also be considered. To that end a finite element mesh of the punch and of the half-space, as shown in Fig. 4.2, is used. The mesh modelling the half-space was generated from a set of elliptic coordinates. The numerical integration of the boundary integrals, as described in Section 3, is performed using the trapezoidal quadrature rule. As seen in Fig. 4.2 the finite element nodes on the contact boundary occur as opposite pairs and the integration points of the quadrature rule are chosen to coincide with these pairs of nodes. In the calculations below punch rigidity is simulated using a Young's modulus 10^5 times larger than that for the half-space, which in all cases were 2000.

In Ref. [8] nondimensional contact stresses and displacements were introduced by the following transformations:

$$\frac{p_N(1-\nu)a}{G\delta} \rightarrow p_N, \quad \frac{p_T(1-\nu)a}{G\delta} \rightarrow p_T, \quad \frac{w_N}{\delta} \rightarrow w_N, \quad \frac{w_T}{\delta} \rightarrow w_T,$$

where G is the shear modulus and ν is Poisson's ratio. It was seen that, when using Coulomb's law of friction in a problem with a rigid punch, the nondimensional solution depends on ν and the coefficient of friction μ, only. When the finite element method is used a finite "half-space" has to be treated; the solution will then depend also on a and δ. However, for a sufficiently large "half-space" this dependency will be small and a unit punch radius is therefore treated below. Furthermore, when our more general friciton law is considered it can be seen that the nondimensionalization implies the following transformations of the contact stiffnesses and the hardening modulus:

$$\frac{Q_N(1-\nu)a}{G} \rightarrow Q_N, \quad \frac{Q_T(1-\nu)a}{G} \rightarrow Q_T, \quad \frac{H(1-\nu)a}{G} \rightarrow H.$$

Therefore, the nondimensional solution will in this case depend on the punch radius and on the ratios of contact stiffnesses and hardening modulus to $G/(1-\nu)$ as well as on ν and μ. Finally, it is also clear that if an elastic punch is treated the solution will depend on the ratio between punch and half-space elastic moduli.

Due to the impossibility of modelling an infinite half-space, the compliances obtained, i. e. the indentation per unit applied force P, will differ from the ones obtained by Spence, even under conditions that are identical in other aspects. In Fig. 4.3 this effect is investigated for the cases of a rigid punch and for $\nu = 0$. A dimensionless compliance factor $P^* = \frac{(1-\nu)P}{4Ga\delta}$ is evaluated for different radii of adhesion c. These radii correspond to different friction coefficients as shown by Spence.

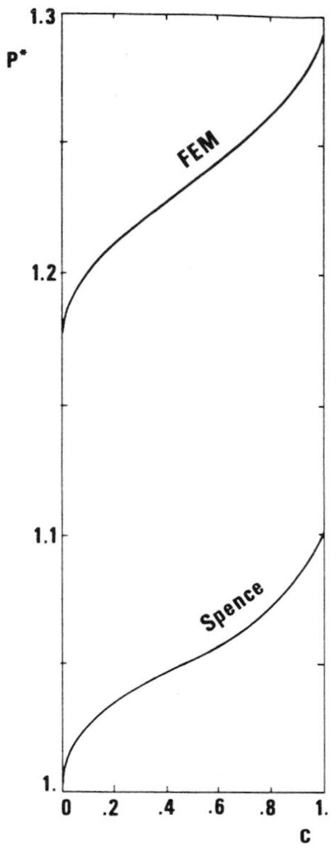

Fig. 4.3 Compliance factor P^*.

In order to investigate how closely the original problems is approximated by the finite element discretization the frictionless case, for which an analytical solution exists, was considered. It was seen that for a prescribed load (i. e. the results in the case of prescribed displacements was scaled by a factor obtained by considering the compliance) there was a close agreement between the general shapes of the analytical and the numerical curves except for the punch edge where the analytical solution is infinite. However, the numerical values were generally 2 % larger than the analytical ones. This can be explained by the presence of the singularity in the frictionless case since this singularity carries part of the load. A similar tendency seems to exist in the numerical solutions by Spence.

The influence of hardening on the contact stress distribution can be estimated by studying Fig. 4.4. These calculations are carried out for $\nu = 0.3$ and $\mu = 0.1843$ which in the case of no hardening results in $c = 0.7$, a value used in the numerical calculations by both Turner and Torstenfelt. Fig. 4.4 shows the whole sequence of loading and unloading and the relative friction coefficient $p_T/\mu p_N$ is there plotted against the radius for different values of hardening and of the unloading parameter β. It is seen that the overall behavior of the solution is not changed by hardening. The drop of the relative friction coefficient near $r = 1$ for $\beta = 1.$ is believed to be due to the singularity of p_N at this radius. The somewhat strange behavior observed for the reversed slip region at $\beta = .2449$ is due to the physically unreal behavior of the kinematic hardening model when $p_N \to 0$. This behavior becomes even more accentuated as $\beta \to 0$.

In Figs. 4.5 and 4.6 the dependency of interface compliance on P^* and c is investigated. These calculations are carried out for $\nu = 0.3$ and $\mu = 0.1843$; the ratio between Q_T and Q_N is kept constant at 0.8. In Fig. 4.7 the stress distribution for two different interface compliances are compared with the curves valid for a rigid interface. The accentuated decrease of normal contact stress near $r = 1.$, when compliance is present, should be noted.

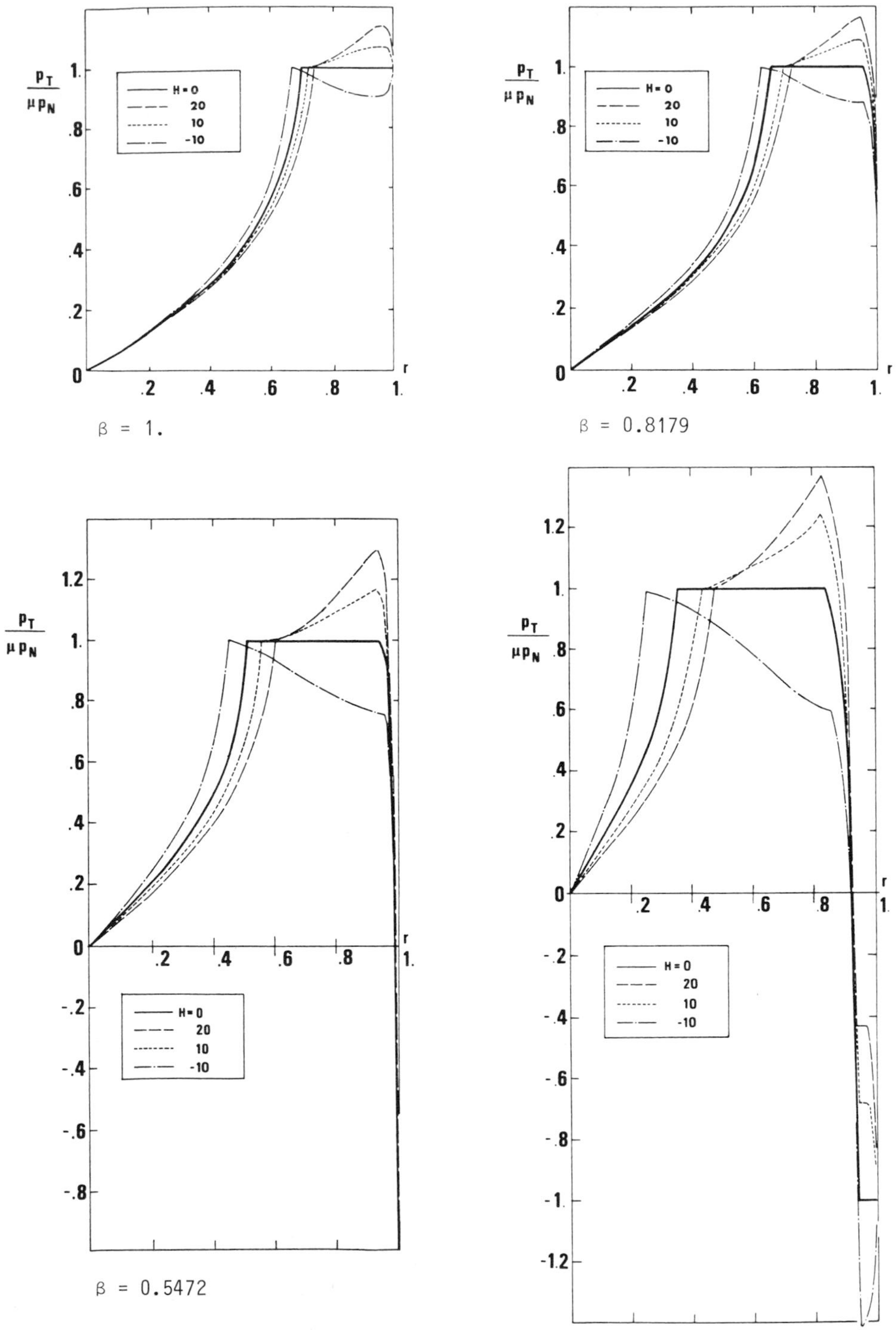

β = 0.2449

Fig. 4.4 The influence of hardening.

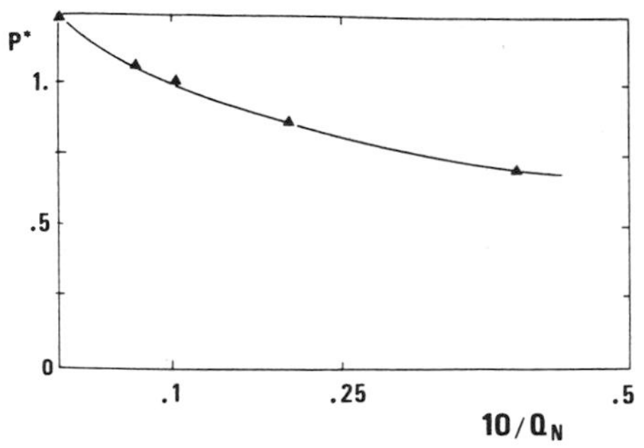

Fig. 4.5 Dependence of compliance factor on interface compliance.

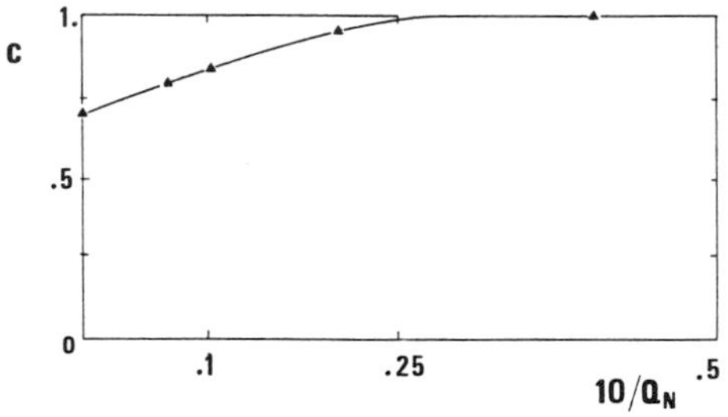

Fig. 4.6 Dependence of radius of adhesion on interface compliance.

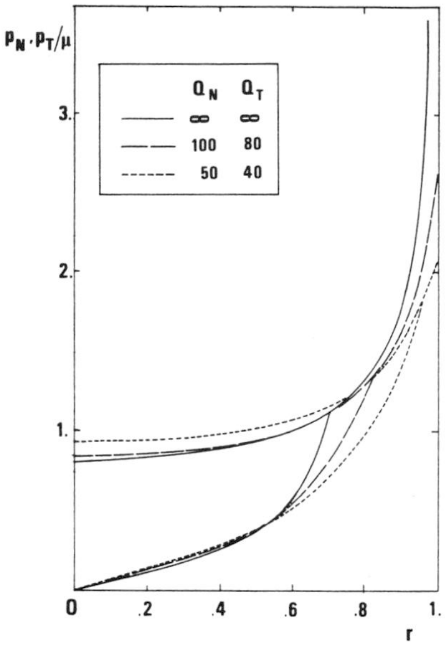

Fig. 4.7 Influence of interface compliance on contact stress distributions.

The influence of non-rigidity of the punch is shown in Figs. 4.8-4.10; $\mu = 0.1843$ and $\nu = 0.3$ for both punch and half-space. It is seen that the radius of adhesion depends on the elasticity of the punch. However, for soft punches a constant value is approached and the qualitative results are expected to agree with the ones obtained in Ref. [22] for a rectangel compressed by rigid planes.

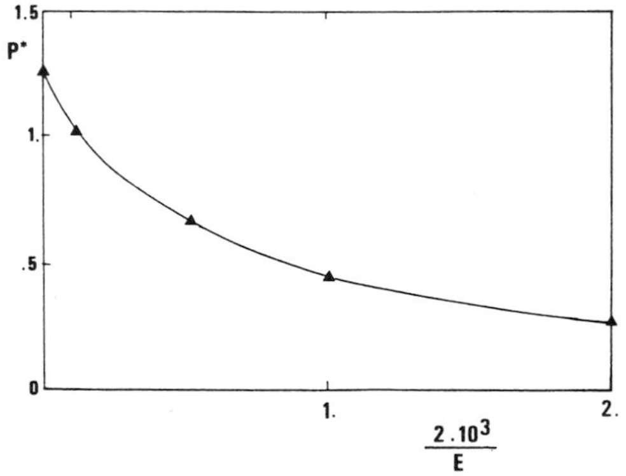

Fig. 4.8 Dependence of compliance factor on elastic modulus of punch.

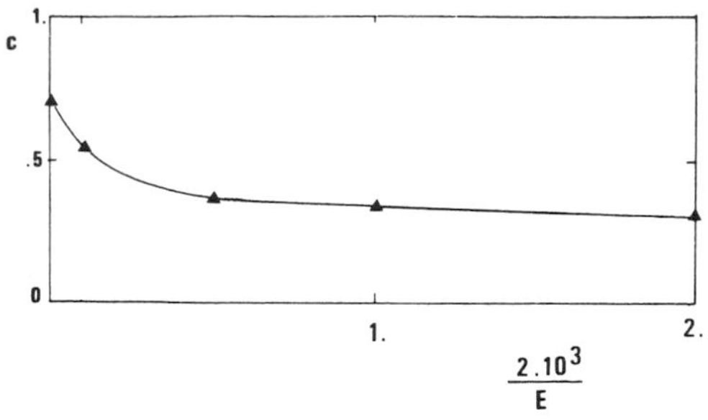

Fig. 4.9 Dependence of radius of adhesion on elastic modulus of punch.

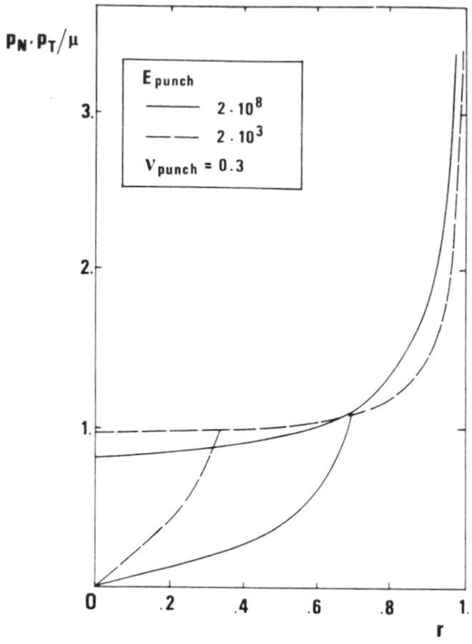

Fig. 4.10 Influence of non-rigidity of punch on contact stress distributions.

REFERENCES

1 J.T. Oden, E.B. Pires, Nonlocal and Nonlinear Friction Laws and Variational Principles for Contact Problems in Elasticity, J. of Appl. Mech., 50 (1983) 67-76.
2 A. Curnier, A Theory of Friction, Int. J. Solids Struct., 20, 7 (1984) 637-647.
3 B. Fredriksson, Finite Element Solution of Surface Nonlinearities in Structural Mechanics with Special Emphasis to Contact and Fracture Mechanics Problems, Comp. & Struct., 6 (1976) 281-290.
4 R. Michalowski, Z. Mróz, Associated and Non-associated Sliding Rules in Contact Friction Problems, Archives of Mech., Archiwum Mechaniki Stosowanej. 30, 3 (1978) 259-276.
5 A. Klarbring, General Contact Boundary Conditions and the Analysis of Frictional Systems. To appear.
6 D.E. Grierson et al, Mathematical Programming and Nonlinear Finite Element Analysis, Comp. Meth. in Appl. Mech. and Eng., 17/18 (1979) 497-518.
7 I. Kaneko, A Parametric Linear Complementarity Problem Involving Derivatives, Mathematical Programming, 15 (1978) 146-154.
8 D.A. Spence, The Hertz Contact Problem with Finite Friction, J. of Elasticity, 5, 3-4 (November 1975).
9 J.R. Turner, The Frictional Unloading Problem on a Linear Elastic Half-space, J. Inst. Maths. Applics., 24 (1979) 439-469.
10 B.R. Torstenfelt, An Automatic Incrementation Technique for Contact Problems with Friction, Comp. & Struct., 19, 3 (1984) 393-400.
11 N. Back, M. Burdekin and A.Cowley, Review of the Research on Fixed and Sliding Joints, Proceedings 13th International Machine Tool Design and Research Conference, London 1973, pp 87-99, S.A. Tobias and F. Koenigsberger (Eds.).

12 J.T. Oden and J.A.C. Martins, Models and Computational Methods for Dynamic Friction Phenomena. To appear: Comp. Meth. Appl. Mech. Eng. (1985).
13 A. Taniguchi, M. Tsutsumi and Y. Ito, Treatment of Contact Stiffness in Structural Analysis, Bulletin of JSME, 27, 225 (March 1984), 601-607.
14 K.L. Johnson, Surface Interaction Between Elastically Loaded Bodies under Tangential Forces, Proc. Roy. Soc., (1955) 531-548.
15 J.S. Courtney-Pratt and E. Eisner, The Effect of a Tangential Force on the Contact of Metallic Bodies, Proceedings of the Royal Society of London, Series A238 (1957).
16 M. Lindgren, Drehmoment-Übertragung in Pressverbindungen, Konstruktion, 25, Heft 9 (1973) 338-341.
17 M. Burdekin, A. Cowley, N. Back, An Elastic Mechanism for the Micro-sliding Characteristics Between Contacting Machined Surfaces, J. Mech. Eng. Science, 20, 3 (1978) 121-127.
18 L. Demkowicz, J.T. Oden, On Some Existence and Uniqueness Results in Contact Problems with Nonlocal Friciton, Nonlinear Analysis, Theory, Methods & Appl., 6, 10 (1982) 1075-1093.
19 J.T. Oden and N. Kikuchi, Finite Element Methods for Constrained Problems in Elasticity, Int. J. for Num. Meth. in Eng., 18 (1982) 701-725.
20 A. Klarbring, A Mathematical Programming Approach to Threedimensional Contact Problems with Friction. To appear.
21 D.A. Spence, A Mathematical Model of Frictional Contact, Proceedings of Euromech Colloquium No. 110 on Contact Problems and Load Transfer in Mechanical Assemblages, Rimforsa, Sweden (1978), B.G.A. Persson, B. Fredriksson and L. Bolin (Eds.).
22 S.N. Prasad and S. Dasgupta, Effect of Sliding Friction on Contact Stresses in a Rectangle Compressed by Rigid Planes, J. of Appl. Mech., 42 (1975) 656-662.

MODELLING OF INTERFACES IN GEOMECHANICS

ON THE MODELING OF CONTACT PROBLEMS WITH DILATION

M. PLESHA[1] and T. BELYTSCHKO[2]

[1]Department of Engineering Mechanics, University of Wisconsin, Madison, Wisconsin 53706 (USA)

[2]Department of Civil Engineering, Northwestern University, Evanston, Illinois 60201 (USA)

ABSTRACT

Constitutive models for joints are reviewed with particular emphasis on those which model dilation. Some recent work on constitutive equations which utilize an elastic-plastic framework and an asperity model with degradation is summarized and some typical results are given.

INTRODUCTION

Contact-friction is an important phenomenon in the behavior of jointed rock masses and structure-media interaction problems. The oldest and most popular idealization for friction is due to Coulomb and, although originally postulated for contact between rigid bodies, has an analogous interpretation for contact between deformable bodies and can be stated as follows: relative motion between two adjacent points in contact will occur when the tangential stress attains a value equal in magnitude to the normal compressive stress multiplied by a constant called the friction coefficient which characterizes the interface. Furthermore, it requires the relative tangential motion be zero prior to frictional slip.

Because of its inherent simplicity and the rather poor understanding of frictional phenomena, Coulomb's law has been almost universally employed for the idealization of friction in the analysis of contact problems. There is, however, considerable experimental evidence which indicates that reversible tangential displacements take place prior to any permanent frictional sliding, i.e. that the interaction forces are elastic and that the initial configuration is sometimes restored if forces are released. Evidence of this behavior has been reported in machined metallic surface contact problems. In some situations, evidence suggests that the macroscopic coefficient of friction increases with increasing displacement; see for example Johnson (ref.1) and Bowden and Tabor (ref.2). Such behavior is attributable to the fact that actual contact takes place at only a few asperity peaks and that the true contact area is considerably smaller than the apparent or macroscopic contact area; a lucid account of the microscopic behavior of these asperities is given by Oden and Pires (ref.3).

Many geotechnical contact problems, such as rock joint problems, are also characterized by reversible tangential displacements prior to frictional sliding; see for example Goodman, Heuze and Ohnishi (ref.4) and Rosso (ref.5). Unlike metallic contact problems, however, in which the contact surfaces have unmatched irregularities which were created by an abrasive machining process, geotechnical contact problems frequently have fully mated (or matched) contact surfaces which are generated by a natural process of cracking and perhaps mineral deposition. Because of the almost fully-mated and rough nature of these contacts, tangential displacements are generally accompanied by normal contact surface displacements; e.g., dilation. Furthermore, because tangential sliding produces asperity fracturing and rubbelization, the amount of dilation and frictional behavior of an interface change with the evolution of the deformation.

A number of investigators have developed models for this behavior, but none of these can be considered completely satisfactory. Patton (ref.6,7) proposed a model consisting of an interface represented by asperity teeth oriented at an angle with respect to the mean plane of the contact. For "low" compressive stresses, the model behavior is characterized by dilation and the overriding of asperities. For "high" stresses, the model is characterized by shearing through asperities. A similar model was proposed by Jaeger (ref.8) which featured a smooth transition from asperity overriding to asperity shearing; Ladanyi and Archambault (ref.9) presented a model with an asperity strength. Barton (ref.10) proposed an empirical model based on a joint roughness coefficient. Roberds and Einstein (ref.11) developed a constitutive law by considering elastic-plastic deformations of an interface and Goodman (ref.12) and Brown (ref.13) developed models by idealizing an interface to consist of nonlinear springs; also see Heuze and Barbour (ref.14).

The Patton, Jaeger and Ladanyi and Archambault models strive to model the physical phenomena of underlying joint behavior, and while they describe the shape of the failure envelope, they do not account for presliding behavior. Furthermore, they do not provide an explicit relationship between increments of displacement and increments of stress, which is necessary for any constitutive law that is to be used in simulation software. The Barton model is a useful empirical tool for shear strength criteria, however, it also does not provide a relationship between displacements and stresses.

The Roberds and Einstein model represents a significant advance in the modeling of dilation in that the phenomenology of presliding elastic behavior and postsliding plastic behavior are considered. However, explicit relationships between increments of stress and deformation were not presented and it is questionable if such relations are obtainable considering the complexity of their theory. The Goodman model describes presliding and postsliding behavior through the use of nonlinear springs or contact stiffnesses. In practice, however, an iterative scheme must be employed to determine the correct spring coefficients so that the desired constitutive law is simulated. The model is not based on the underlying physics of contact behavior and is only applicable for monotonic loading.

Dilation, asperity shearing and rubbelization are difficult effects to account for without the aid of a general and consistent framework. Such a framework has been partially introduced by Fredriksson (ref.15) and Michalowski and Mróz (ref.16). In this paper, we will examine constitutive relations for contact-friction within the framework of a plasticity theory in the spirit of (ref.15-16). Special attention will be given to the development of constitutive relations for contact-friction with dilation, i.e. frictional slip which is accompanied by contact volume expansion. This is an important effect in many geotechnical contact problems which, to date, has not received adequate attention.

CONSTITUTIVE EQUATIONS FOR JOINTS

We consider a generic, two-dimensional joint which separates two contacting bodies B_1 and B_2 as shown in Fig. 1. The nomenclature is defined in the right-hand side of the figure, which shows the normal and tangential directions t and n. The relative tangential and normal deformations, g_t and g_n respectively, are given by

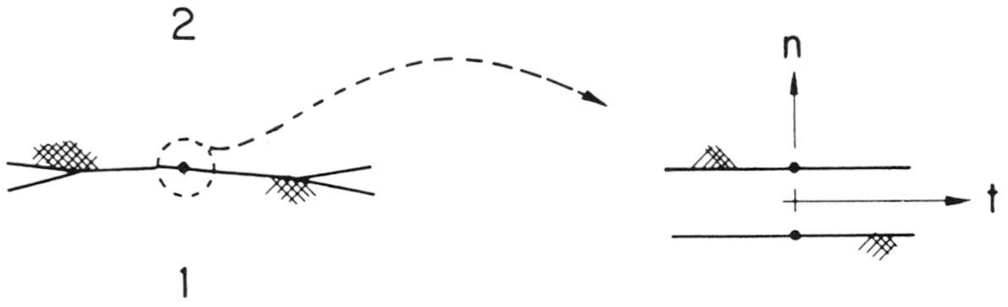

Fig. 1. Nomenclature for contact surface description.

$$g_t = (\underset{\sim}{u}_2 - \underset{\sim}{u}_1) \cdot \hat{\underset{\sim}{t}}$$
$$g_n = (\underset{\sim}{u}_2 - \underset{\sim}{u}_1) \cdot \hat{\underset{\sim}{n}} \qquad (2.1)$$

where $\underset{\sim}{u}_1$ and $\underset{\sim}{u}_2$ are the displacements of the contact point associated with bodies 1 and 2 and $\hat{\underset{\sim}{t}}$ and $\hat{\underset{\sim}{n}}$ are unit vectors in the positive tangential and normal directions.

Although the bodies are considered in contact, relative normal deformation g_n is allowed to take place while the bodies remain in contact in an effort to model the behavior of asperities and any low strength interfacial layer such as mineral deposits. Thus, in contrast to the classical contact problem, a positive value of g_n does not necessarily imply loss of contact because dilatory behavior of the joint can lead to separation without causing the normal stress to vanish. Associated with these directions are the interface tractions, or contact stresses σ_t and σ_n. The objective of the constitutive law is the development of a relation between the forces across the joint and the macroscopic deformations g_t and g_n which reflects any microstructural features of the interface such as, for example, matched or unmatched asperity teeth.

In a constitutive approach to contact modeling, the interface layer is considered to carry normal and tangential stresses σ_n and σ_t (in complete notation the stresses would be denoted by σ_{nn} and σ_{nt}), within an infinitesimal layer of material separating the two bodies. The normal stress (σ_{tt} in complete notation) is ignored. If elasto-plasticity is then used as a framework for this constitutive law, the rates of deformation are additively decomposed into a recoverable elastic part and a nonrecoverable plastic part

$$\dot{g}_i = \dot{g}_i^e + \dot{g}_i^p \qquad (2.2)$$

and the rate of the interface stress is given by the elastic constitutive law

$$\dot{\sigma}_i = E_{ij}\, \dot{g}_j^e \qquad (2.3)$$

where E_{ij} are the interface stiffnesses. In this notation superscripts e and p denote elastic and plastic parts, respectively, a superposed dot denotes a time derivative and subscripts indicate vector components; repeated indices denote summation unless otherwise stated.

The constitutive model, like any nonassociated flow model, involves a yield function $F(\underset{\sim}{\sigma}, \underset{\sim}{g})$ and a slip potential $G(\underset{\sim}{\sigma}, \underset{\sim}{g})$ whose gradient defines the plastic slip. The yield function defines the onset of plastic deformation: plastic deformation occurs if and only if F is zero and remains zero; whenever $\dot{F}$ is negative, yielding ceases and the behavior becomes elastic.

Thus the plastic rate of deformation is given by the flow rule as follows

$$\dot{g}_i^p = \begin{cases} 0 & \text{if} \quad F < 0 \quad \text{or} \quad \dot{F} < 0 \\ \dot{\lambda} \dfrac{\partial G}{\partial \sigma_i} & \text{if} \quad F = 0 \quad \text{and} \quad \dot{F} = 0 \end{cases} \tag{2.4}$$

When F and G are identical functions, the flow rule is associated and when they are different the flow rule is called nonassociated.

Embodied in Eqs. (2.2) through (2.4) is the framework for a general contact-friction constitutive law. The slip function F and the slip potential G can be selected to reflect the microstructural features of the contact. Despite the great similarity between friction and plasticity, Drucker (ref.17) demonstrated that a friction law as simple as Coulomb's cannot be expressed by an associated plastic flow rule (where the plastic rate of deformation is normal to the yield, or slip surface) because this would require that the contacting bodies have, at the onset of friction, a separating velocity in the normal direction, which is not present in Coulomb's law. Therefore, friction is, in general, nonassociated and F and G are different. Only in the very special case of contact in which a thick seam of a plastic material separates the two surfaces can the friction be associative; under these circumstances the interface can support a normal separating velocity during frictional sliding.

SLIP CRITERIA AND SLIP POTENTIALS FOR CONTACT WITH DILATION

Of the little literature which has appeared on the subject of slip criteria and slip potentials for friction, almost all has been devoted to the description of smooth contact problems such as machined metallic contact problems which are nondilating; see Fredriksson (ref.15) and Curnier (ref.18). However, Michalowski and Mróz (ref.16) present slip criteria and flow rules for a Patton type model (ref.6,7) consisting of an interface with fully mated asperity teeth as shown in Fig. 2. In this work, a Coulomb friction law is assumed on each asperity surface and the slip criteria and flow rules are derived in terms of the macrosopic interface stresses and deformations.

These models have been extended in the work of Plesha (ref.19) which will be summarized here. It is based on a Patton type model in which the model contact problem shown in Fig. 2 is assumed to have matched asperity teeth as microstructural features. In addition, it is assumed that a mechanism exists for asperity degradation, e.g., asperity shearing.

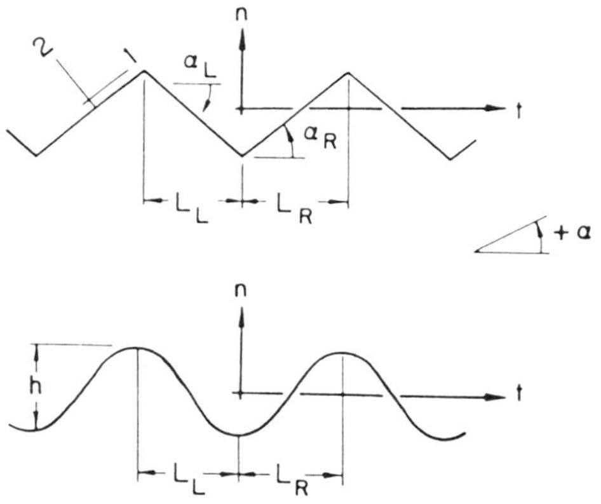

Fig. 2. Asperity models and nomenclature.

The development of the constitutive relation parallels that employed in the mathematical theory of plasticity (ref. 20). If slip is imminent then the slip function vanishes, $F_K = 0$; here K denotes which asperity flank is active, i.e, slip to the left or right. Furthermore, if at the next instant in time we remain on the yield, or slip surface, then

$$\dot{F}_K = 0 \tag{3.1}$$

If we restrict attention to slip functions such that $F_K = F_K(\underset{\sim}{\sigma})$ and if we assume that the degradation of the asperity angle is a function of tangential plastic work, W_t^p, only, then (3.1) becomes

$$\frac{\partial F_K}{\partial \sigma_i} \dot{\sigma}_i + \frac{\partial F_K}{\partial \alpha_K} \frac{\partial \alpha_K}{\partial W_t^p} \dot{W}_t^p = 0 \qquad \text{no sum on } K \tag{3.2}$$

Combining (3.2) with (2.2) through (2.4) and noting that $\dot{W}_t^p = \sigma_t \, \dot{g}_t^p$ provides the incremental contact-friction constitutive relation

$$\dot{\sigma}_i = E_{ij}^{ep} \dot{g}_j \tag{3.3}$$

where

$$E_{ij}^{ep} = E_{ij} - \frac{\frac{\partial F_K}{\partial \sigma_p} E_{pj} E_{iq} \frac{\partial G_K}{\partial \sigma_q}}{\frac{\partial F_K}{\partial \sigma_p} E_{pq} \frac{\partial G_K}{\partial \sigma_q} - H} \qquad \text{no sum on } K \tag{3.4a}$$

and

$$H = \frac{\partial F_K}{\partial \alpha_K} \frac{\partial \alpha_K}{\partial W_t^p} \sigma_t \frac{\partial G_K}{\partial \sigma_t} \tag{3.4b}$$

We wish to emphasize that Eqs. (3.3-4) is an explicit relationship between rates (or increments) of stress and rates (or increments) of displacement which is easily evaluated and can be easily implemented in numerical procedures.

This framework can be used to include standard Coulomb friction with a smooth contact surface. The yield function in this case is

$$F = (\sigma_t^2)^{1/2} + \mu \, \sigma_n \tag{3.5}$$

and the potential function is

$$G = (\sigma_t^2)^{1/2} \tag{3.6}$$

If we assume $E_{tn} = E_{nt} = H = 0$, then (3.4-6) give

$$\dot{\sigma}_t = -\mu E_n \operatorname{sgn}(\sigma_t) \dot{g}_n \tag{3.7a}$$

$$\dot{\sigma}_n = E_n \dot{g}_n \tag{3.7b}$$

where $\operatorname{sgn}(\sigma_t)$ denote the sign of σ_t. Equations (3.7) can be combined to give the standard relation for increments of stress in Coulomb friction without reversal, namely

$$\dot{\sigma}_t = -\mu \operatorname{sgn}(\sigma_t) \dot{\sigma}_n \tag{3.8}$$

The dilatational properties of rock joints can be modeled by incorporating a microstructural contact surface idealization. The surface is idealized to consist of identical asperities with a size which is characteristic of the most predominant features of the discontinuity being modeled. Plesha (ref.19) presented two models for the asperities; these are shown in Fig. 2. The first is a Patton (ref. 6) type model consisting of saw-tooth asperity surfaces which degrade; the second is a sine asperity surface model in which the irregularities are idealized as sine functions which degrade.

For the Patton model, sliding can be active on either the right asperity with angle α_R and mean asperity length L_R, or the left asperity with angle α_L and mean length L_L. The sign convention for asperity angles is positive counterclockwise, thus $\alpha_R > 0$ and $\alpha_L < 0$. For the sine asperity model, sliding can be active on either the right or left asperity surfaces, however, the asperity angle, or in other words, the slope at the tangent point of contact, is a function of the tangential displacement g_t. Letting the asperity height be h, the slope, or instantaneous asperity angle at the tangent point of contact is given by

$$\alpha_K = \frac{\pi h}{2L_K} \sin \frac{\pi}{2} \left(1 + \frac{g_t}{L_K}\right) \tag{3.9}$$

for nonzero g_t which does not exceed the mean asperity length and where subscript K = R if the right asperity is active and K = L if the left asperity is active.

If we assume Coulomb friction on the asperity surface, then, Eqs. (3.5) and (3.6) for the slip function and slip potential are applicable with σ_t and σ_n replaced by σ_1 and σ_2 respectively, where 1 and 2 denote the tangent and normal directions to the active asperity surface. By simple transformation, these stresses can be expressed in the macroscopic (t, n) coordinate system of the interface which results in

$$F = [(\sigma_n \sin \alpha_K + \sigma_t \cos \alpha_K)^2]^{1/2} + \mu(\sigma_n \cos \alpha_K - \sigma_t \sin \alpha_K) \qquad (3.10)$$

$$G = [(\sigma_n \sin \alpha_K + \sigma_t \cos \alpha_K)^2]^{1/2} + \text{constant} \qquad (3.11)$$

and α_K can be taken from either the Patton asperity model or the sine asperity model and K denotes which asperity surface is active. The friction coefficient μ is taken to be equal to the tangent of the residual friction angle of the joint, which is readily measurable.

The degradation of the asperities is also an important consideration. The asperity behavior we wish to model is characterized as follows: under high compressive stresses, high tangential stresses are required to produce slip, and rapid asperity degradation can occur for small tangential displacements. Under low compressive stresses, low tangential stresses will produce slip, yet if the amount of the slip is large, then asperity degradation can arise from surface wear. A simple model for asperity degradation which is capable of replicating the salient features of this type of asperity wear and accounts for the irreversible nature of damage is obtained by assuming that degradation is a function of the plastic tangential work W_t^p; $\dot{W}_t^p = \sigma_t \dot{g}_t^p$. Thus, if a particular degree of degradation is obtained by a state of high stress and low displacement, the same degree of degradation can be obtained at a lower stress level provided the displacement is sufficiently large. The relatively simple tribological relationship we will employ is given by

$$\alpha_K = (\alpha_K)_o \exp(-cW_t^p) \qquad (3.12)$$

where $(\alpha_K)_o$ is the initial asperity surface angle and c is a rock joint degradation constant which has units of length/force and reflects how rapidly the asperity surfaces deteriorate.

While a number of assumptions have been made in these developments, the examples of the next section show that the model is in qualitative and quantitative agreement with experimental observations and commonly held notions regarding joint behavior. Nevertheless, there are a number of open questions which require additional research.

EXAMPLES

To illustrate the constitutive law developed in this paper we will numerically simulate a direct shear test with a prescribed normal displacement for three rock types consisting of "weak rock," "moderately strong" rock and "strong" rock. The model is shown in Fig. 3 and consists of a rock interface with a coefficient of friction of $\mu = 1/2$ and an initial asperity orientation of 8° with respect to the horizontal. The joint has elastic stiffnesses $E_t = 0.5\ GN/m^3$ and $E_n = 1.0\ GN/m^3$ and is initially compressed such that the compressive stress has a magnitude of 1 MPa . For the duration of the simulation, the change in normal displacement is prescribed to be zero, hence, dilation is constrained. The model is loaded by prescribing the tangential motion of the upper plate of material. Loading is unidirectional and cyclic behavior is not considered although it is an important aspect of the model behavior which requires investigation.

The strength of the asperity teeth is accounted for through the selection of the asperity degradation constant, c in Eq. (3.12). Infinitely strong asperity teeth correspond to c = 0 and increasingly weaker teeth correspond to an increasing degradation constant. In the examples which follow, "strong," "moderately strong" and "weak" rock were simulated by using asperity degradation coefficients of 0.01, 0.3 and 1.0 m/kN respectively.

The solution consists of the normal and tangential stress histories as shown in Fig. 3 for the 3 rock types. The broken curve illustrates the tangential stress history for a typical planar nondilating Coulomb type interface. The results show good qualitative agreements with typical experimental observations; see Goodman (ref.21), pg. 164. Because of the constrained dilation, the normal stress is increasing for all cases. The weaker rocks, which experience more rapid asperity shearing, develop lower compressive stresses and after small amounts of displacement, show no additional increases.

The tangential stress results indicate some interesting behavior which occurs in real rock joints. The strongest rock considered demonstrates a uniform increase in tangential stress because of the presence of increasing

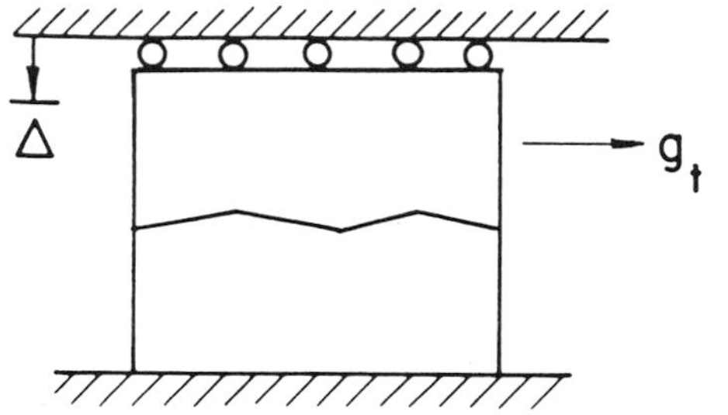

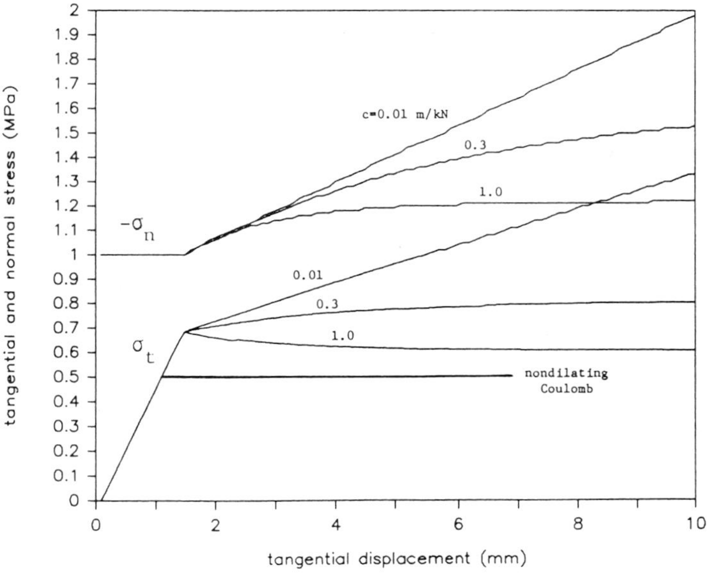

Fig. 3. Results for Plesha model (ref.19) with different rock strengths.

(a)
(b)
(c)

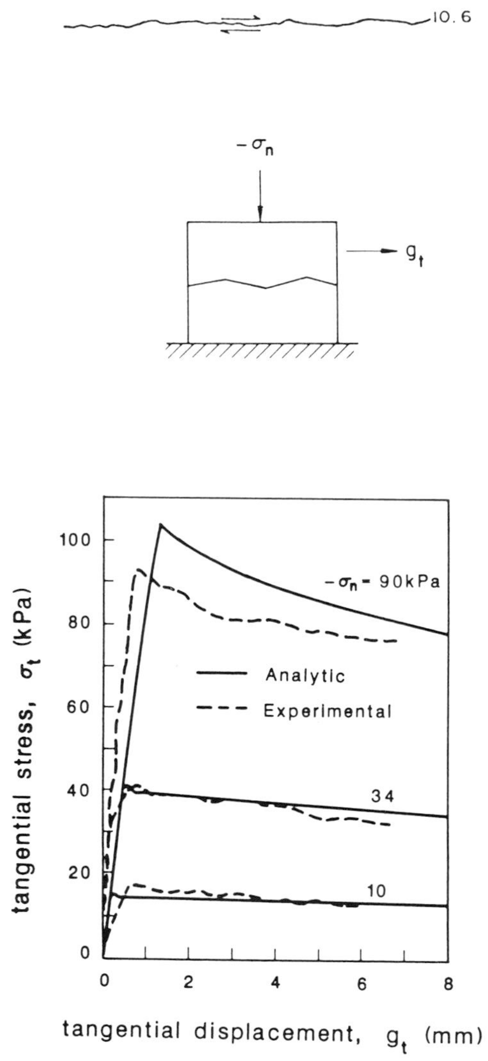

Fig. 4. Comparison of Plesha model (ref.19) with experiment.

normal stress. The moderately strong rock, however, displays only a very small variation after the initiation of sliding despite the presence of a moderately increasing normal stress. The physical interpretation of this phenomenon is that while the increasing compressive stress contributes to increased tangential stress, the asperities are degrading at such a rate that the overall effect is compensating. The weak rock, after initial slipping demonstrates a marked decrease in tangential stress, again in the presence of moderately increasing compressive stress. The physical interpretation of this phenomenon is that the asperities are degrading so rapidly that the joint weakens with increasing tangential displacement. While this may appear to be unstable behavior because of the negative tangent modulus, the total system may be stable because of the release of blocked elastic energy; compare Bazant (ref. 22).

The last example (ref.19) compares the predictions of this model with the tests performed in (ref.23). The surface profile of the joint is shown in Fig. 4a; the joint roughness and residual friction angle are reported as 10.6 and 32°, respectively. The experiments employed constant normal stresses of 10, 34 and 90 kPa and prescribed shear displacements of 6 to 7 mm. Based on the experimental results, $E_t = 0.08$ GN/m^3 and $\mu = 0.625$; E_n was arbitrarily chosen as 1.0 GN/m^3; a discussion of the effect of joint stiffness and a comprehensive tabulation is given in Belytschko et al. (ref.24). Based on the joint profile, α was estimated to be 16°.

The model and experimental results are compared in Fig. 4c for a monotonic shearing. The agreement is quite good both qualitatively and quantitatively.

ACKNOWLEDGEMENT

We gratefully acknowledge the support of the National Science Foundation under Grant CEE-8314440.

REFERENCES

1. K.L. Johnson, Surface Interaction Between Elastically Loaded Bodies Under Tangential Forces, Proc. Royal Society, Series A, 230 (1955) 531-548.
2. F.P. Bowden and D. Tabor, The Friction and Lubrication of Solids, Part II, Clarendon Press, Oxford, 1964.
3. J.T. Oden and E.B. Pires, Nonlocal and Nonlinear Friction Laws and Variational Principles for Contact Problems in Elasticity, J. Applied Mechanics, 50 (1983) 67-76.
4. R.E. Goodman, F.E. Heuze and Y. Ohnishi, Research on Strength-Deformability-Water Pressure Relationship for Faults in Direct Shear, Final Report on ARPA Contract H0210020, University of California, Berkeley, 1972.
5. R. Rosso, A Comparison of Joint Stiffness Measurements in Direct Shear, Triaxial Compression and In-Situ, Int'l. J. Rock. Mech. Min. Sci. and Geomech. Abst., 13 (1976) 167-172.

6. F.D. Patton, Multiple Modes of Shear Failure in Rock and Related Materials, Ph.D. Thesis, University of Illinois, 1966.

7. F.D. Patton, Multiple Modes of Shear Failure in Rock, Proc. 1st Cong. ISRM, Lisbon, 1, 509-513.

8. J.C. Jaeger, Friction of Rocks and the Stability of Rock Slopes-Rankine Lecture, Geotechnique, 21 (1971) 97-134.

9. B. Ladanyi and G. Archambault, Simulation of the Shear Behavior of Jointed Rock Mass, Proc. 11th Symposium on Rock Mech., (1970) 105-125.

10. N. Barton, Review of a New Shear Strength Criterion for Rock Joints, Engrg. Geology, 7 (1973) 287-332.

11. W.J. Roberds and H.H. Einstein, Comprehensive Model for Rock Discontinuities, J. Geotechnical Engr., ASCE, 104 (1978) 553-569.

12. R.E. Goodman, Methods of Geological Engineering in Discontinuous Rocks, West Publishing Co., St. Paul, MN, 1976.

13. D.M. Brown, Numerical Simulations of Earthquake Effects on Tunnels for Generic Nuclear Waste Repositories, Report of Science Applications, Inc., SAI-FR-126, Vol. II, Appendix C (1980) 1-44.

14. F.E. Heuze and T.G. Barbour, New Models for Rock Joints and Interfaces, J. Geotechnical Engineering Division, ASCE, 108 (1982) 757-776.

15. B. Fredriksson, Finite Element Solution of Surface Nonlinearities in Structural Mechanics with Special Emphasis to Contact and Fracture Mechanics Problems, Computers and Structures, 6 (1976) 281-290.

16. R. Michalowski, and Z. Mroz, Associated and Non-Associated Sliding Rules in Contact Friction Problems, Archives of Mechanics, 30 (1978) 259-276.

17. D.C. Drucker, Coulomb Friction, Plasticity and Limit Loads, Journal of Applied Mechanics, ASME, 21 (1954) 71-74.

18. A. Curnier, A Theory of Friction, Int'l. J. Solids and Structures, 20 (1984) 637-647.

19. M. Plesha, Constitutive Models for Rock Discontinuities with Dilatancy and Surface Degradation, submitted for publication.

20. R. Hill, The Mathematical Theory of Plasticity, University Press, Oxford, 1960.

21. R.E. Goodman, Introduction to Rock Mechanics, John Wiley and Sons, New York, 1980.

22. Z.P. Bazant, Work Inequalities for Plastic Fracturing Materials, Int'l J. Solids and Structures, 16 (1980) 873-901.

23. S. Bandis, A.C. Lumsden and N.R. Barton, Experimental Studies of Scale Effects on the Shear Behavior of Rock Joints, Int'l J. Rock Mech. Min. Sci. and Geomech. Abs., 18 (1981) 1-21.

24. T. Belytschko, M.E. Plesha and C.H. Dowding, A Computer Method for Stability Analysis of Caverns in Jointed Rock, Int'l. J. Num. and Anal. Meth. in Geomech., 8 (1984) 473-492.

TORSIONAL RESPONSE OF CYLINDRICAL RIGID INCLUSIONS WITH INTERFACE SLIP

A.P.S. SELVADURAI and R.K.N.D. RAJAPAKSE
Department of Civil Engineering, Carleton University, Ottawa, Ontario, Canada
K1S 5B6

ABSTRACT

Classical studies related to the problem of the load transfer between a rigid cylindrical inclusion and an elastic halfspace assume perfect bonded contact at the interface. The present paper re-examines the load transfer problem in the light of an interface response which is elastic-perfectly plastic. The non-linear relationship between the applied torque and the corresponding rotation of the cylindrical inclusion is evaluated by employing an analytical-numerical scheme which discretizes the interface shear stress distribution.

INTRODUCTION

The group of problems which examine the behaviour of cylindrical inclusions which are partially or fully embedded in elastic media have several useful engineering applications. In the context of geomechanics, the fully embedded rigid inclusion serves as an elementary model of an anchoring region which is used to provide reaction against gravitational and uplift loads. Similarly, the problem of the cylindrical rigid inclusion embedded at the surface of a halfspace models the behaviour of a pile or pier embedded in a geological medium. In the context of mechanics of composite materials, the cylindrical inclusion problem can be used to model the local action of multiphase materials which are reinforced with low concentrations of rod-like fibres. The model of the interaction between the inclusion and the surrounding elastic medium can also be used to investigate flaw bridging and dowel action that takes place at fractured interfaces of composite and other reinforced materials. These applications are schematically illustrated in Fig. 1.

A number of researchers have investigated various groups of problems related to inclusions embedded in elastic media. The fundamental analytical work of Muki and Sternberg (ref.1) examines the problem related to the load transfer from a cylindrical elastic inclusion into an elastic infinite space. In this study the elastic bar with properties different from the surrounding elastic infinite space was subjected to uniform traction over its cross section. The decay of the resultant force in the bar was determined by an analytical solution of the governing integro-differential equation system. These authors

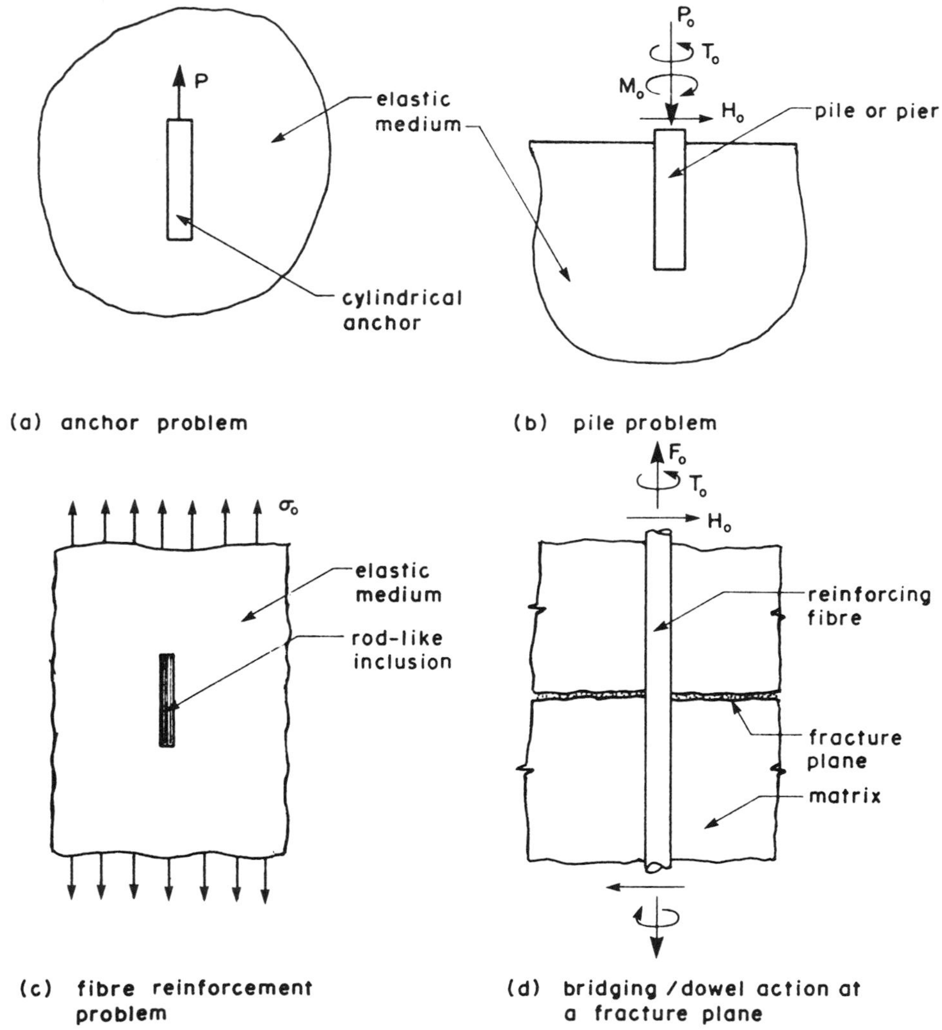

Figure 1: Engineering applications of the cylindrical inclusion model.

also indicated an approximate procedure which can be used to examine the decay of load in a bar with an arbitrary cross-section. In a subsequent paper Muki and Sternberg (ref. 2) extended the work to include load transfer from a cylindrical bar into an isotropic elastic halfspace. The assumption of a uniform stress distribution across the cross section of the bar essentially reduces the problem to the solution of a Fredholm integral equation of the second-kind. The solution procedure developed by Muki and Sternberg (ref. 2) is valid for situations in which the length/diameter ratio of the bar is comparatively large. An exact analytical formulation of the problem of a rigid cylindrical inclusion which is embedded at the surface of an isotropic elastic halfspace was given by Luk and Keer (ref. 3). A Hankel transform development of the governing equations yields a system of coupled singular integral equations for the normal and shear traction distributions at the inclusion elastic medium interface. The procedures used for the formulation, analysis and the numerical computation are rigorous. The work of Luk and Keer (ref. 3) can be regarded as the complete solution to the problem of the axially loaded cylindrical inclusion which is embedded at the surface of a halfspace. In an earlier paper, Keer and Freeman (ref. 4) examined the problem of the transfer of torque from a semi-infinite elastic inclusion into an elastic halfspace. Owing to the spatial symmetry of the torsional loading, the solution to the problem related to the halfspace also furnishes the solution to the torsional loading of a cylindrical bar of infinite length embedded in an elastic infinite space. A number of other researchers have investigated the load transfer problem by incorporating analytical techniques of varying rigour. For example, Spillers and Stoll (ref. 5) considered the problem of the lateral loading of an elastic pile which is embedded in an elastic halfspace. The pile is modelled as a line inclusion which obeys the Bernoulli-Euler beam theory and the interactive reaction between the beam and the halfspace is modelled by the horizontal analogue of the Mindlin force problem (ref. 6). A major limitation of this modelling scheme is the absence of vertical shear stresses on the cylindrical surface, normal stresses at the base and the lack of compatibility of vertical displacement between the elastic halfspace and the embedded bar. Problems relating to the axial loading of incompressible piles embedded in elastic media were examined by Poulos and Davis (refs. 7,8). These authors also invoke an approximation scheme similar to that used by Spillers and Stoll (ref.5). The pile-soil interface was discretized into a number of ring elements of finite length. Expressions for the axial displacement due to vertical stresses acting on the discrete regions were derived by employing Mindlin's fundamental solution (ref. 9) for the concentrated axial load acting at the interior of a halfspace. The intensity of the interface stresses acting on each elemental region was determined by prescribing the compatibility of displacements between the pier and the elastic halfspace at the interface. Suriyamongkol et al. (ref.10) examined the behaviour of an axially loaded rigid cylindrical body embedded in bonded contact at the surface of an elastic halfspace. The region of the embedded body and the remainder of the halfspace were treated as a single domain and the field of distributed traction resultants were applied at the boundary of the cylindrical region. Again, the requisite axial and radial compatibility conditions are used to determine the intensities of the traction resultants at the discrete locations. The investigations by Niumpradit and Karasudhi (ref.11) and Apirathvorakij and Karasudhi (ref.12) extended the analytical scheme proposed by Muki and Sternberg (ref.2) to include quasistatic axial load transfer and bending problems for a cylindrical elastic bar which is embedded in a fluid-saturated poroelastic halfspace. The basic analytical scheme proposed by Muki and Sternberg (ref.2) has also been used by Fowler and Sinclair (ref.13) to study the elastodynamic problem of the cylindrical bar which is embedded in an elastic halfspace. The elastostatic problem of an elastic bar of infinite length which is embedded in an elastic infinite space and subjected to an infinite train of periodic axial forces was considered by Parnes (ref.14). Herrmann et al. (ref.15) also investigated the transfer of load from a microfibre into an elastic medium, in connection with their studies on the buckling of individual fibres located in elastic materials with low relative rigidity. Recent investigations by Selvadurai (ref.16) examine the problem of the lateral loading of a Bernoulli-Euler beam which is embedded in bonded contact with an isotropic elastic infinite space. The accuracy of these analytical estimates have also been verified by appeal to boundary element analyses of the same (ref.17). Selvadurai (ref.18) has also employed the boundary element technique to study the interaction between a flexible structural element such as a pipeline and the surrounding elastic soil, under conditions of ground movement at a fault zone. Luk and Keer (ref.19) has also examined the problem of the axial loading of a cylindrical rigid inclusion which is fully embedded in bonded contact within an isotropic elastic infinite space. Again, a Hankel transform development of the governing equations is used to generate the coupled singular integral equation system. A numerical evaluation of the system is used to generate the load-displacement response of the anchor. These results compare favourably with exact analytical results derived by Selvadurai (ref.20) for the axial loading of rigid spheroidal inclusions with an identical geometric aspect ratio. Recently, Rajapakse and Selvadurai (ref.21) and Selvadurai and Rajapakse (ref.22) have examined the class of problems related to the generalized loading of uniform and non-uniform cylindrical rigid inclusion (both solid and hollow) which are embedded in non-homogeneous and layered elastic media. The above review of the elasticity problem relating to the load transfer from an embedded inclusion into an elastic halfspace is by no means extensive; references to further work are given by Sternberg (ref.23) and Gladwell (ref.24).

An examination of the literature on load transfer problems in classical elasticity indicates that in all investigations (excepting ref.7), the interface is assumed to exhibit a perfectly bonded contact at all stress levels. In a variety of engineering situations, however, this assumption is not rigorously satisfied. For example in the case of a pile embedded in a geological medium the interface can exhibit a non-linear response consistent with the mechanical properties of the geological medium and the method of installation of the pile. It is therefore desirable to re-examine the load transfer problem in the light of a non-linear response of the interface. The basic assumption in the ensuing analysis is that the non-linearity is local and restricted only to the interface region. The halfspace or the infinite space which contains the cylindrical inclusion exhibits an elastic response. The paper will focus on the torsional loading of a rigid cylindrical pile which is embedded at the surface of an elastic halfspace (Fig. 2). The interface is assumed to exhibit an elastic-perfectly plastic response in shear (Fig. 3a). During the application of the torque, regions of the interface will undergo yielding. Slippage will occur in the yielded regions and bonded contact is maintained in the unyielded regions. As the rotation is increased there is progressive yield at the interface. A rigorous analytical solution of this class of load transfer problems involving non-linear interfaces is quite complicated. For this reason the paper focusses on the use of discretization techniques (refs. 7,21,22) to examine the influence of interface yielding on the torsional load transfer problem. Numerical results presented in the paper illustrate the manner in which the torque vs. rotation relationship for the embedded inclusion is influenced by the length/diameter aspect ratio of the inclusion and the interface shear strength. The methodology outlined in the paper can be extended to the examination of other forms of interface non-linearities (Fig. 3b).

FUNDAMENTAL SOLUTION

The discretization technique used to analyse the torsional load transfer from the inclusion to the elastic medium (either with/or without non-linear interface responses) commences with subdivision of the interface region into segments of uniform stress. The segment dimensions can be chosen to reflect the plausible stress gradients that can be encountered in a particular inclusion configuration. In order to determine the displacement field associated with each region of uniform traction it is necessary to develop an appropriate fundamental solution. These fundamental solutions can take into account appropriate symmetries that are associated with global deformations of the cylindrical inclusion. In the case of the torsional loading of the inclusion

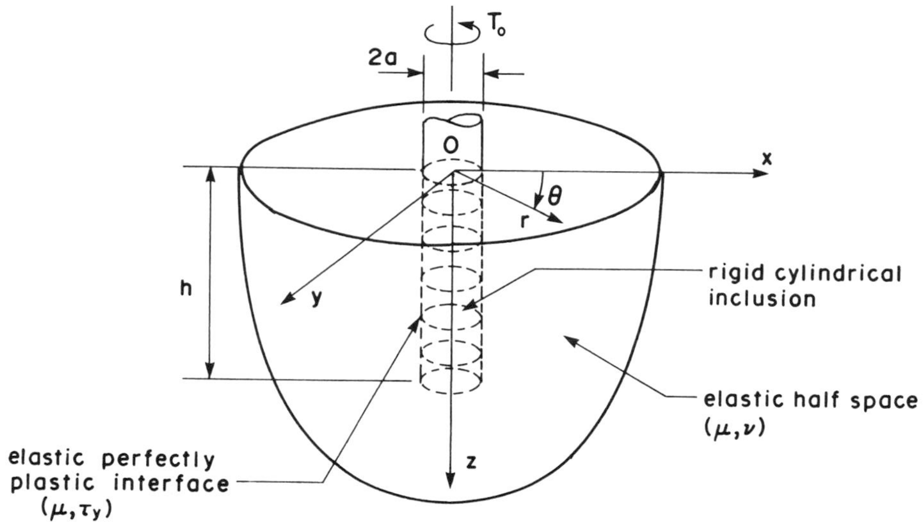

Figure 2: Geometry of the bar and the discretization scheme for the inclusion-elastic-medium interface.

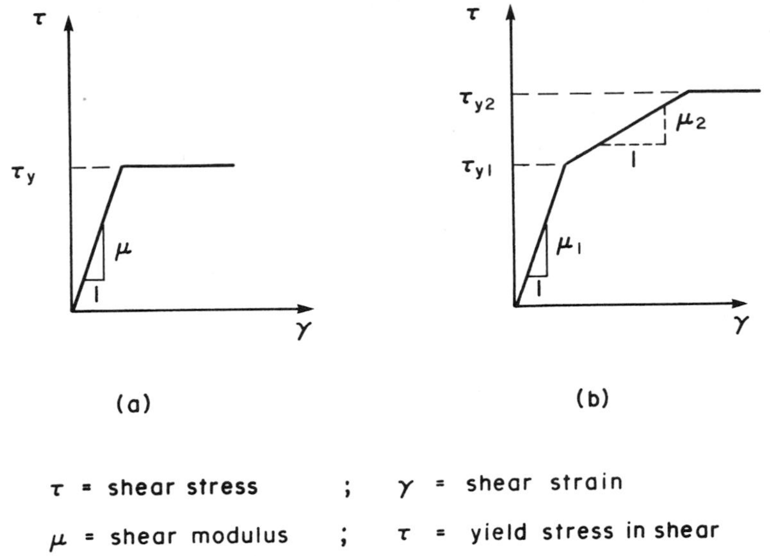

Figure 3. Non-linear interface responses.

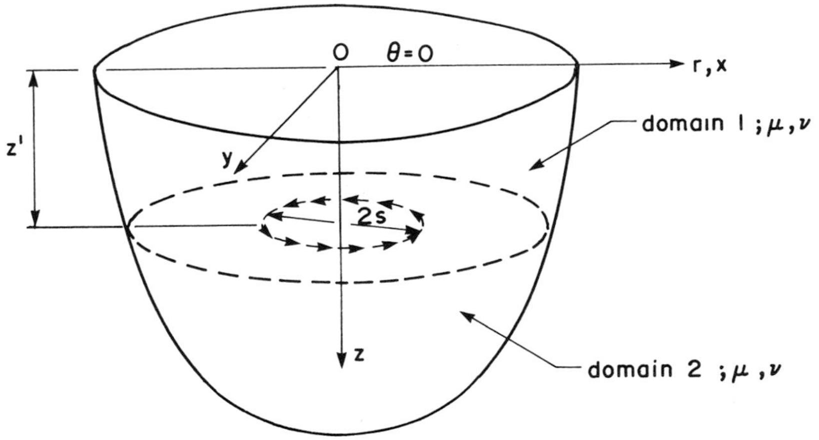

Figure 4. Sub-regions considered in the derivation of the fundamental solution.

the deformation exhibits a state of axisymmetric torsion about the z-axis. Therefore the deformation field is characterized by

$$u_r = 0 \; ; \; u_\theta = v(r,z) \; ; \; u_z = 0 \tag{1}$$

where (u_r, u_θ, u_z) are the components of the displacement vector referred to the cylindrical polar coordinate system (r,θ,z). The non-zero components of the stress field referred to the cylindrical polar coordinate system are

$$\underset{\sim}{\sigma} = \begin{bmatrix} 0 & \sigma_{r\theta} & 0 \\ \sigma_{r\theta} & 0 & \sigma_{\theta z} \\ 0 & \sigma_{\theta z} & 0 \end{bmatrix} \tag{2}$$

In the absence of body forces, the non-trivial displacement equation of equilibrium is given by

$$\frac{\partial^2 v}{\partial r^2} + \frac{1}{r}\frac{\partial v}{\partial r} + \frac{\partial^2 v}{\partial z^2} - \frac{v}{r^2} = 0 \tag{3}$$

The global solution of (3) applicable to regions in which $r \varepsilon (0,\infty)$ can be obtained by employing a Hankel transform development of the same (ref. 25). It can be shown that

$$v(r,z) = \int_0^\infty [A(\xi)e^{-\xi z} + B(\xi)e^{\xi z}] \, J_1(\xi r)d\xi \tag{4}$$

where $A(\xi)$ and $B(\xi)$ are arbitrary functions which are to be determined by adopting the appropriate boundary conditions. We now adopt this general result to develop a fundamental solution to the problem of the internal loading of a halfspace region by a concentrated ring load of unit intensity located at a depth $z=z'$ below the free surface (Fig. 4). For the purpose of this analysis we identify two domains as follows: domain 1 in which $r \varepsilon (0,\infty)$; $z \varepsilon (0,z')$ and domain 2 in which $r \varepsilon (0,\infty)$; $z \varepsilon (z',\infty)$. The general solutions for the displacement field in each domain is identified by the notations $v^{(1)}(r,z)$ and $v^{(2)}(r,z)$. The relevant integral expressions for the displacement fields are

$$v^{(1)}(r,z) = \int_0^\infty [A^{(1)}(\xi)e^{-\xi z} + B^{(1)}(\xi)e^{\xi z}] \, J_1(\xi r)d\xi \tag{5}$$

and

$$v^{(2)}(r,z) = \int_0^\infty [A^{(2)}(\xi)e^{-\xi z} + B^{(2)}(\xi)e^{\xi z}] \, J_1(\xi r)d\xi \tag{6}$$

Also, the displacement fields and the stresses derived from the displacement fields must exhibit regular behaviour as $(r^2+z^2)^{1/2} \to \infty$. In this case, the function $B^{(2)}(\xi)$ must be set equal to zero. The remaining functions $A^{(1)}(\xi)$, $A^{(2)}(\xi)$ and $B^{(1)}(\xi)$ can be determined from the traction boundary conditions at z=0 and the continuity conditions at z=z', i.e.

$$\sigma_{\theta z}^{(1)}(r,0) = 0 \tag{7}$$

$$v^{(1)}(r,z') = v^{(2)}(r,z') \tag{8}$$

$$\sigma_{\theta z}^{(1)}(r,z') - \sigma_{\theta z}^{(2)}(r,z') = \delta(r-s) = \int_0^{\infty} s\,\xi\, J_1(\xi s)\, J_1(\xi r)d\xi \tag{9}$$

and $\delta(r-s)$ is the delta function at r=s. Avoiding details of calculations it can be shown that the final solutions for the displacement fields $v^{(1)}(r,z)$ and $v^{(2)}(r,z)$ take the forms

$$v^{(1)}(r,z) = \frac{1}{2\mu}\int_0^{\infty} s\, J_1(\xi s)\, J_1(\xi r)\,[e^{-\xi(z+z')} + e^{-\xi(z'-z)}]d\xi \quad ;\ 0 \le z \le z' \tag{10a}$$

and

$$v^{(2)}(r,z) = \frac{1}{2\mu}\int_0^{\infty} s\, J_1(\xi s)\, J_1(\xi r)[e^{-\xi(z+z')} + e^{-\xi(z-z')}]d\xi \quad ;\ z' \le z < \infty \tag{10b}$$

respectively and μ is the linear elastic shear modulus.

UNIFORM RIGID CYLINDRICAL INCLUSION EMBEDDED IN A HALFSPACE

We focus attention on the problem of the torsional loading of a rigid cylindrical inclusion which is embedded in a homogeneous elastic halfspace and subjected to an external torque T_0. It is assumed that prior to yielding at the interface, complete contact is maintained between the inclusion and the surrounding elastic material. In order to evaluate the shear stress distribution along the interface we discretize the axisymmetric interface by regions of uniform or linearly varying shear traction (Fig. 5). It may be observed that the inclusion-elastic medium interface of any cylindrical body can be discretized by using a combination of the two ring elements shown in Fig. 5. These are identified as shaft elements (SE) and base elements (BE). In the case of a shaft element it is assumed that the tractions are distributed uniformly across the thickness of an element. However, the symmetric tangential traction acting on a BE under torsional loading is assumed to have a linear variation across the thickness of an element. The unknown traction acting on the jth element is denoted by T_{mj}^{n} (m=r,θ,z; n=s,b) where the subscript m denotes the direction of traction and the superscript n is used to identify whether

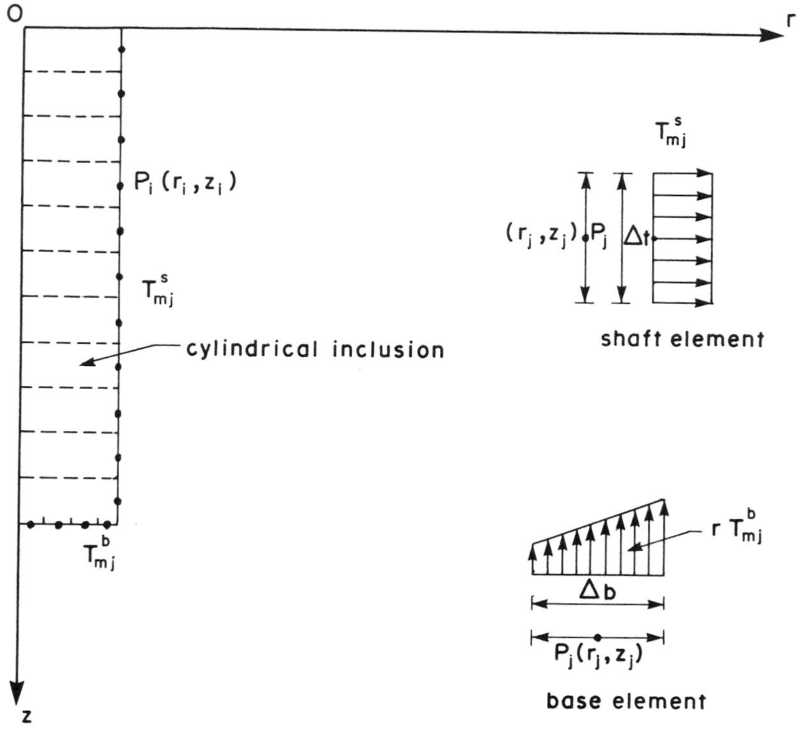

Figure 5. The discretized cylindrical inclusion-elastic medium interface and the basic elements used in the discretization scheme.

the element is a shaft element or base element. The influence function $[f(r_i,z_i,s_j,z_j)]_q^{mn}$ $(m,q=r,\theta,z;\ n=s,b)$ denotes the displacement in the m-direction at a point $P_i(r_i,z_i)$ due to a unit traction acting in the q-direction on the jth element. The superscript n is used to identify whether the jth element is a shaft element or a base element. The expressions for $[f(r_i,z_i,s_j,z_j)]_q^{mn}$ are obtained by integrating the fundamental solutions, given by equations (10), across the thickness of the corresponding element. For the torsional loading

$$[f(r_i,z_i,s_j,z_j)]_\theta^{\theta s} = \frac{s_j}{2\mu}\,[\alpha I_1(1,1;-1)-I_2(1,1;-1)]\Big|_{z'=z_j-\frac{\Delta t}{2}}^{z'=z_j+\frac{\Delta t}{2}} \tag{11a}$$

$$[f(r_i,z_i,s_j,z_j)]_\theta^{\theta b} = \frac{1}{2\mu}\,[I_1^*(1,2;-1)+I_2^*(1,2;-1)]\Big|_{s=s_j-\frac{\Delta b}{2}}^{s=s_j+\frac{\Delta b}{2}} \tag{11b}$$

where Δt and Δb are the thickness of the shaft and base elements respectively. In equations (11a) and (11b)

$$\alpha = -1 \text{ if } z_i < z_j - \Delta t/2$$

$$\alpha = 1 \text{ if } z_i > z_j + \Delta t/2 \tag{12}$$

If $z_j-\Delta t/2 < z_i < z_j + \Delta t/2$, then the influence functions given by (11a) and (11b) should be computed by considering the integration over the intervals $[z_j-\Delta t/2,\ z_j+\Delta t/2]$ as summation of integration over the intervals $[z_j-\Delta t/2, z_i]$ and $[z_i, z_j+\ t/2]$ with the appropriate value of α given by (12). Also in equations (11a) and (11b)

$$I_1(m,n;p) = \int_0^\infty J_m(\xi r_i)J_n(\xi s_j)\xi^p\, e^{-\xi|z_i-z'|}\,d\xi \tag{13}$$

$$I_2(m,n;p) = \int_0^\infty J_m(\xi r_i)J_n(\xi s_j)\xi^p\, e^{-\xi|z_i+z'|}\,d\xi \tag{14}$$

$$I_1^*(m,n;p) = \int_0^\infty J_m(\xi r_i)J_n(\xi s)\,\xi^p\, e^{-\xi|z_j-z_i|}\,d\xi \tag{15}$$

$$I_2^*(m,n;p) = \int_0^\infty J_m(\xi r_i)J_n(\xi s)\xi^p\, e^{-\xi|z_j+z_i|}\,d\xi \tag{16}$$

The flexibility equation for the rigid cylindrical inclusion can be established in the following manner. The displacements $v_i(r_i,z_i)$ for torsional displacement at a point $P_i(r_i,z_i)$ due to the traction field $T_{\theta j}^{\ s}$ or $T_{\theta j}^{\ b}$ at the interface can be expressed as

$$v_i(r_i,z_i) = \sum_{j=1}^{N_s} [f(r_i,z_i,s_j,z_j)]_\theta^{\theta s} T_{\theta j}^{\ s} + \sum_{j=1}^{N_b} [f(r_i,z_i,s_j,z_j)]_\theta^{\theta b} T_{\theta j}^{\ b} \tag{17}$$

where N_s and N_b are the number of shaft and base elements respectively. For a rigid cylinder under a torsional rotation ϕ_0 we have

$$v_i(r_i,z_i) = \phi_0 \, r_i \tag{18}$$

Considering the compatibility of displacements between the rigid cylinder and the elastic halfspace at the interface, we obtain a system of simultaneous equations ($i=1,2,\ldots N_s+N_b$) for the unknown tractions $T_{\theta j}^{\ n}$ corresponding to each discretized region. The relationship between the rotation ϕ_0 and the applied torque T_0 can be obtained by considering the global equilibrium of the inclusion which is expressed in terms of the interface tractions $T_{\theta j}^{\ n}$, i.e.

$$T_0 = 2\pi \sum_{j=1}^{N_s} r_j^{\,2} \, T_{\theta j}^{\ s} \, \Delta t + \pi \sum_{j=1}^{N_b} r_j^{\,3} \, T_{\theta j}^{\ b} \, \Delta b \tag{19}$$

The preceding analysis focusses only the situation where perfect bonding exists at the inclusion-elastic medium. In the case where the interface is capable of yielding, perfect contact is maintained for all values of inclusion rotations for which $T_{\theta j}^{\ n} < \tau_y$. When the interface traction at a node reaches the limiting shear stress τ_y then slipping occurs at that particular node. If we account for interface behaviour with slipping, then the analytical procedure has to be performed in an incremental manner. Initially, values are selected for the incremental rotations $\delta\phi_0$ of the rigid cylindrical inclusion. The flexibility equations given by (18) are solved for the incremental interface tractions. At each increment, the numerical values of the total tractions at the interface are examined to establish yielded and unyielded interface regions. On each yielded node at the interface, the tractions are known quantities which are determined from the interface shear strength (τ_y). The compatibility condition is also relaxed at a yielded node (i.e., the row and column corresponding to that element is removed from the flexibility equation for the system). The displacements of elements which have not yielded are corrected to accommodate contributions from the elements which have yielded. For each increment of rotation ($\delta\phi_0$) this procedure is repeated until stable regions of yielded or unyielded regions of the interface are identified. The procedure is repeated with the next increment of rotation. The procedure is repeated until all the elements at the interface region experiences yielding. For the completely yielded interface, the ultimate torque $T_0^{\,y}$ can be determined from purely equilibrium considerations, i.e.

$$T_0^y = \{2\pi a^2 h + \frac{2\pi a^3}{3}\} \tau_y \qquad (20)$$

NUMERICAL RESULTS

The computation of the influence function $[f(r_i,z_i,s_j,z_j)]_q^{mn}$ appearing in the flexibility equation is discussed in (refs. 21 and 22). An important factor in the computation is the selection of the values of increments $\delta\phi_0$. If the value of $\delta\phi_0$ is too large, then the ultimate value of T_0^y will be larger than that given by (20). This is due to the fact that within an increment tractions can increase considerably beyond the limiting value. However, this discrepancy can be corrected by checking the difference between the maximum value of traction and the limiting value and reducing the rotation increment for the current step by linear proportioning and re-solving the flexibility equations. If the value of $\delta\phi_0$ is such that the traction increment during the rotation increment is about one percent of the limiting value, then the ultimate value obtained from the numerical solution agrees very closely with the result given by (20). Also, it may be noted that the result for the perfectly bonded interface condition agrees with other analytical results given by Rajapakse and Selvadurai (ref. 21), Luco (ref. 26) and Karasudhi et al (ref. 27). The choice of the discretization, as depicted by the values of N_s and N_b, is also very important to the numerical analysis. If the discretization of the contact surface is coarse, then the torque vs rotation relationships derived from that particular discretization scheme deviates considerably from the corresponding result obtained from a refined interface discretization. The numerical analysis is carried out with successive interface discretization refinements until the results for two successive refinements converge approximately to the same result. The Figures 6 and 7 illustrate the torque-twist relationships that are derived for rigid cylindrical inclusions with different length (h) to radius (a) aspect ratios. As h/a increases the non-linear transition region between the initial elastic region and the final ideally plastic region tends to decrease. In the special case when $h \to 0$ the problem corresponds to the "Reissner-Sagoci problem" in which the interface exhibits an elastic perfectly plastic response. The Figure 8 illustrates the manner in which the torque vs rotation relationship varies for a rigid disc which is in contact with an elastic halfspace; the interface non-linearity is characterized by the elastic-perfectly plastic response. Based on the procedures outlined above and the numerical procedures discussed in refs. (21 and 22), the authors have developed a computer code which evaluates, in an iterative manner, the non-linear load-displacement responses for rigid cylindrical inclusions embedded in elastic media and

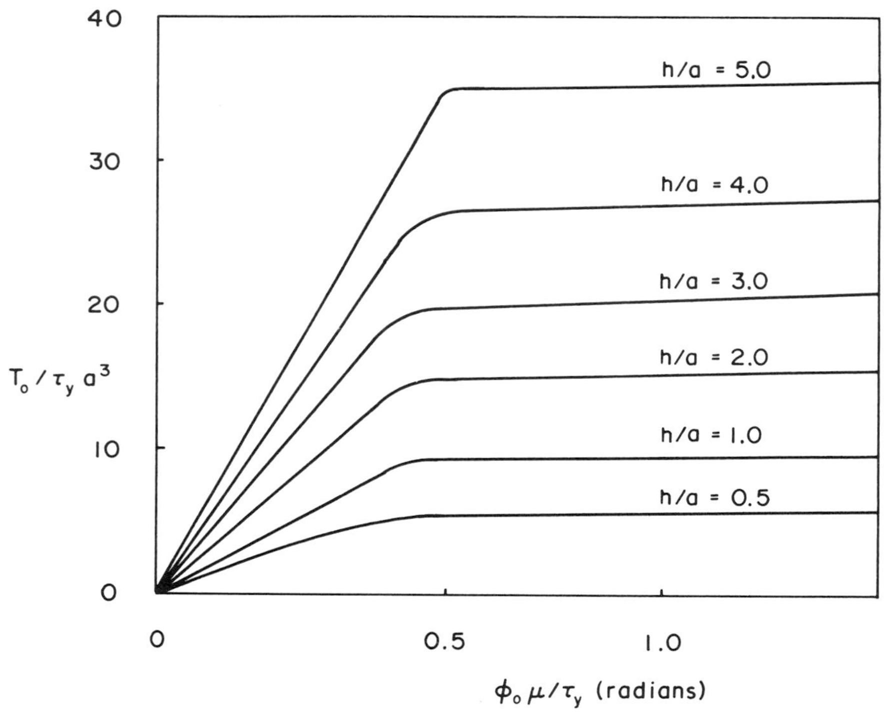

Figure 6. Normalized torque vs. twist relationship for the embedded rigid cylinder.

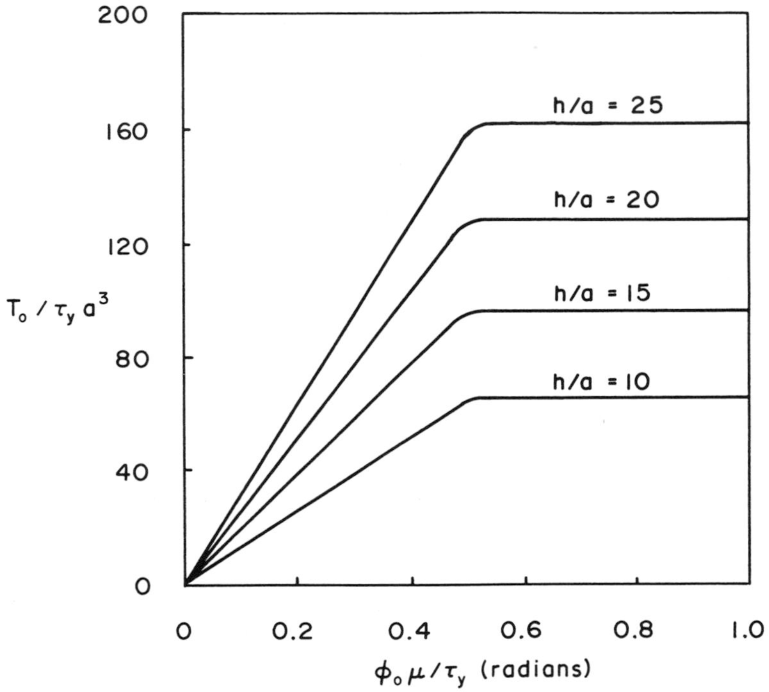

Figure 7. Normalized torque vs. twist relationship for the embedded rigid cylinder.

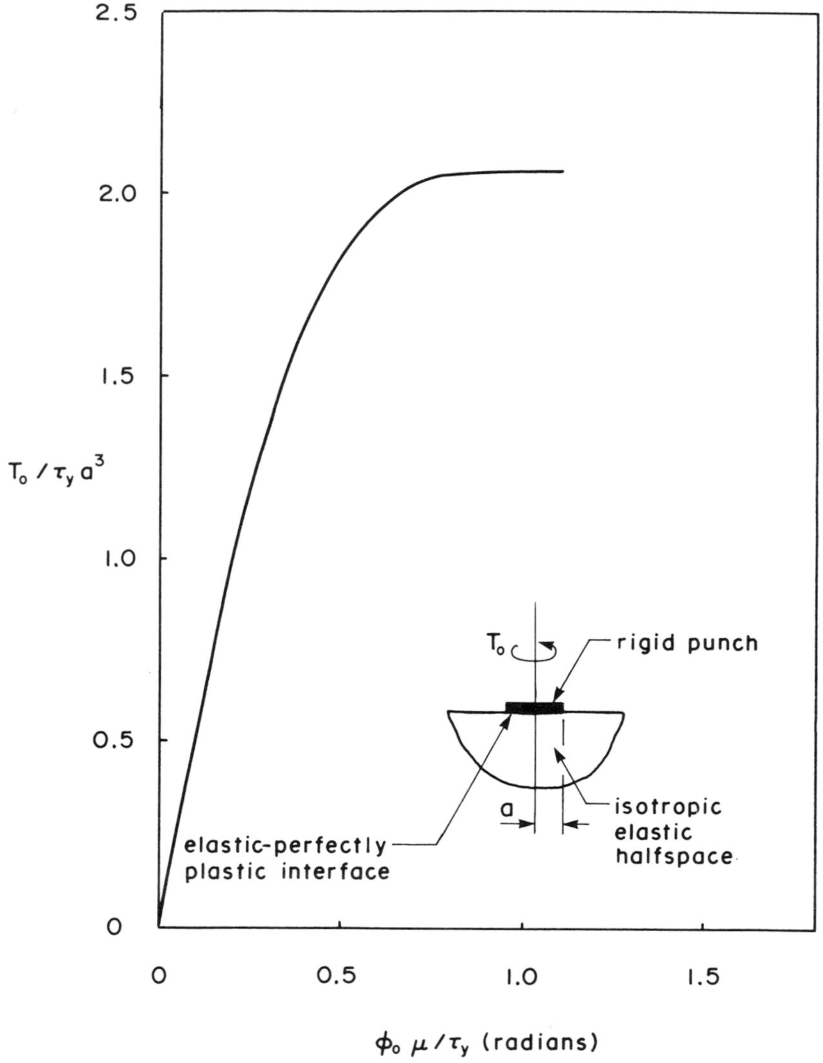

Figure 8. Reissner-Sagoci problem for a rigid circular punch: influence of an elastic-perfectly plastic interface.

possessing an elastic-perfectly plastic interface response. The essentially non-linear code can be easily implemented on a microcomputer. The analysis can also accommodate embedded inclusions which have non-uniform or tapered cross sections; in this case inclined elements are incorporated to account for the non-uniformity.

CONCLUSIONS

The analytical studies pertaining to load transfer from inclusions to an elastic halfspace region focus predominantly on the assumption of a perfectly bonded contact at the interface. The incorporation of interface or local non-linearity can be viewed as simple approximation for the more complicated class of problems in which both the embedded bar and the halfspace can exhibit globally non-linear phenomena. The class of problems which focus on interface non-linear phenomena can of course be examined via a number of numerical techniques such as finite element or boundary element techniques. In this paper we present an alternative numerical scheme in which desirable attributes of an analytical formulation are fully incorporated in the analysis. The resulting discretization technique has been applied to a number of load transfer problems involving cylindrical, tapered and hollow inclusions which exhibit perfect bonded contact. It is shown that the method can be easily extended to incorporate elastic-perfectly plastic phenomena at the interfaces. In the case of long rigid cylindrical inclusions subjected to torsion ($h/a>5$), the transition from an elastic interface response to a completely yielded interface response occurs only over a limited range of rotation (or torque). In this sense the overall torque-rotation response for the embedded cylindrical inclusion can also be approximated by an elastic-perfectly plastic response in which the elastic stiffness and limiting torque are determined separately. In the case of flat disc shaped inclusions, the progressive yielding on the plane faces of the inclusion has a dominant effect on the elastic-plastic transition zone. The numerical scheme proposed in the paper can be further extended to include the problem of the axial tensile loading of cylindrical inclusion embedded at the surface of a halfspace region.

REFERENCES

1 R. Muki and E. Sternberg, Int. J. Solids Struct., 5 (1969) 587-605.
2 R. Muki and E. Sternberg, Int. J. Solids Struct., 6 (1970) 69-87.
3 V.K. Luk and L.M. Keer, Int. J. Solids Struct., 15 (1979) 805-827.
4 L.M. Keer and N.J. Freeman, J. Appl. Mech. 37 (1970) 504-512.
5 W.R. Spillers and R.D. Stoll, J. Soil Mech. Found. Eng. ASCE, 90 (1964) 1-9.
6 R.D. Mindlin and D.H. Cheng, J. Appl. Phys. 21 (1950) 926.
7 H.G. Poulos and E.H. Davis, Geotechnique, 18 (1968) 351-371.
8 H.G. Poulos and E.H. Davis, Pile Foundation Analysis and Design, John Wiley, New York, (1980) 397 pp.

9 R.D. Mindlin, Physics, 7 (1936) 195-202.
10 S. Suriyamongkol, P. Karasudhi and S.L. Lee, Proc. 13th Mid-Western Mechanics Conf., Univ. of Pittsburg, Pa., 1973, 333-347.
11 B. Niumpradit and P. Karasudhi, Int. J. Numer. Anal. Methods Geomech., 5 (1981) 115-138.
12 V. Apirathvorakij and P. Karasudhi, Int. J. Solids Struct., 16 (1980) 625-644.
13 G.F. Fowler and G.B. Sinclair, Int. J. Solids Struct., 14 (1978) 999-1012.
14 R. Parnes, Int. J. Solids Struct., 17 (1981) 891-901.
15 L.R. Herrmann, W.E. Mason and S.T.K. Chan, J. Composite Materials, 1 (1967) 212-226.
16 A.P.S. Selvadurai, Int. J. Numer. Anal. Meth. Geomech. 8 (1984) 157-166.
17 M.C. Au and A.P.S. Selvadurai, Proc. Southeastern Conf. Theor. Appl. Mech., Auburn Univ., XII, 1 (1984) 143-151.
18 A.P.S. Selvadurai, Proc. Int. Conf. on Advances in Underground Pipeline Engineering, Madison, Wisc. (1985) (in press).
19 V.K. Luk and L.M. Keer, Int. J. Numer. Anal. Methods Geomech., 4 (1980) 215-232.
20 A.P.S. Selvadurai, Geotechnique, 26 (1976) 603-612.
21 R.K.N.D. Rajapakse and A.P.S. Selvadurai, Int. J. Numer. Anal. Methods Geomech. (in press).
22 A.P.S. Selvadurai and R.K.N.D. Rajapakse, Int. J. Solids Struct. (in press).
23 E. Sternberg, Proc. 6th U.S. National Congr. of Appl. Mech., Boulder, Co. (1970) 34-61.
24 G.M.L. Gladwell, Contact Problems in the Classical Theory of Elasticity, Sijthoff and Noordhoff, The Netherlands, (1980) 716 pp.
25 I.N. Sneddon, (Ed.) Applications of Integral Transforms in the Theory of Elasticity, CISM Courses and Lectures No.220, Springer Verlag, Wien, 426 pp.
26 J.E. Luco, J. Applied Mechanics, ASME, 43 (1976) 419-423.
27 P. Karasudhi, R.K.N.D. Rajapakse and B.Y. Hwang, Int. J. Solids Struct., 20 (1984) 1-11.

CONSTITUTIVE MODELLING FOR INTERFACES UNDER CYCLIC LOADING

C. S. DESAI and B. K. NAGARAJ
Department of Civil Engineering and Engineering Mechanics, University of Arizona, Tucson, Arizona (USA)

SUMMARY

Importance of appropriate modelling and laboratory testing for constitutive behavior of interfaces in dynamic soil-structure is identified. A brief description of the constitutive models developed by the authors and co-workers is presented here. The models are verified with respect to typical results from special laboratory tests.

INTRODUCTION

The response of structures founded on geologic media, soil or rock (Fig. 1), subjected to static and dynamic loadings such as earthquakes, wave forces and blasts is influenced significantly by the properties of the structural and geologic materials and interfaces/joints. Hence, it is essential to include the effect of nonlinear behavior of geologic materials and interfaces, and also the various relative motions, translation, debonding, rebonding, experienced by the interfaces. When these effects are included, it is difficult to obtain closed-form solutions, and it becomes necessary to use numerical techniques such as the finite element method.

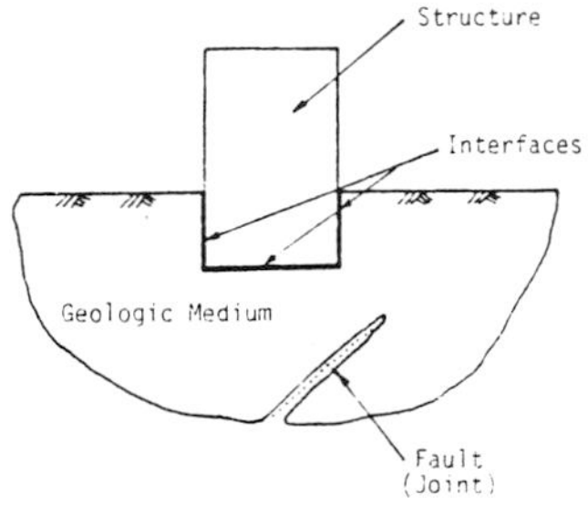

Fig. 1. Typical structure-foundation system

The subject of soil-structure interaction is wide in scope and recently, a number of investigators have considered the foregoing topics. It is not possible to present a detailed review of these works; such reviews are available elsewhere (ref. 1, 2). The main objective of this paper is to present brief details of the recent research performed by the authors and co-workers (ref. 1, 3-10), for modelling of interfaces for nonlinear response.

NONLINEAR BEHAVIOR

Nonlinear response of interfaces is influenced by various factors such as state of stress, stress path, inelasticity, volume changes, type and rate of loading, roughness and modes of deformation (no-slip, slip, debonding and rebonding). Here a model based on piecewise linear behavior is considered and allows for factors such as state of stress, amplitudes, cycles of loading and loading, unloading and reloading.

Interface Model

A rather new and simple model called thin-layer element is proposed to simulate the interface behavior. This model can be based on piecewise nonlinear elastic or elastic-plastic concepts and can incorporate loading, unloading and reloading behavior. It allows for various modes of deformation such as no-slip, slip, debonding or separation, and rebonding, and can control interpenetration at interface (ref. 1, 4, 6). Here only the piecewise nonlinear model is considered, while research on the elastic-plastic concept based on the general model proposed (ref. 11, 12) will be the subject of other publications.

The thin-layer element, Fig. 2, is treated essentially like a solid element with small finite thickness. However, its constitutive behavior is defined differently. The constitutive matrix $[C_i]$ for the interface is defined through incremental stress-strain relation:

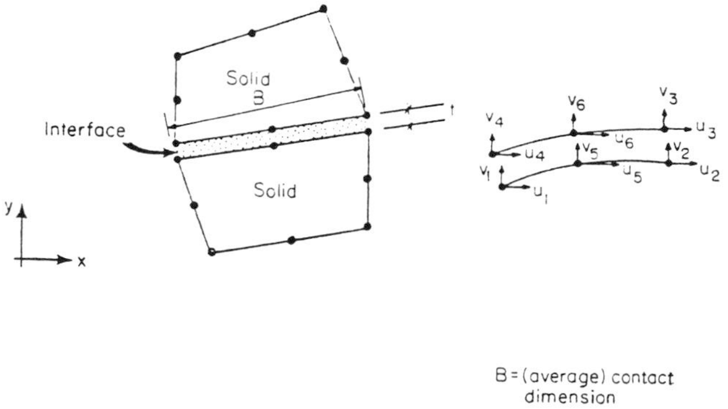

Fig. 2. Two-dimensional solid and thin-layer interface/joint elements

$$\{d\sigma\} = [C_i]\{d\varepsilon\} \tag{1}$$

where

$$[C_i] \quad \begin{bmatrix} [C_n] & [0] \\ [0] & [C_s] \end{bmatrix} \tag{2}$$

where $[C_n]$ and $[C_s]$ are normal and shear components, respectively.

<u>Shear Behavior</u>

For the slip mode with compressive normal stress, the shear part of the stiffness (for nonlinear elastic model) is defined through the shear modulus, G_i, as

$$G_i \cong \frac{\tau}{u_r} t' \tag{3}$$

where τ = shear stress, u_r = relative displacement and t' = (small) thickness. It can be defined as tangent modulus from:

$$\tau = \tau\,(u_r, \sigma_n, \gamma_i, N) \tag{4}$$

where σ_n = normal stress, γ_i = initial density and N = number of loading cycles.

For the nonlinear elastic model, a modified form of the Ramberg-Osgood model (ref. 8, 9, 13, 14) is used; Fig. 3.

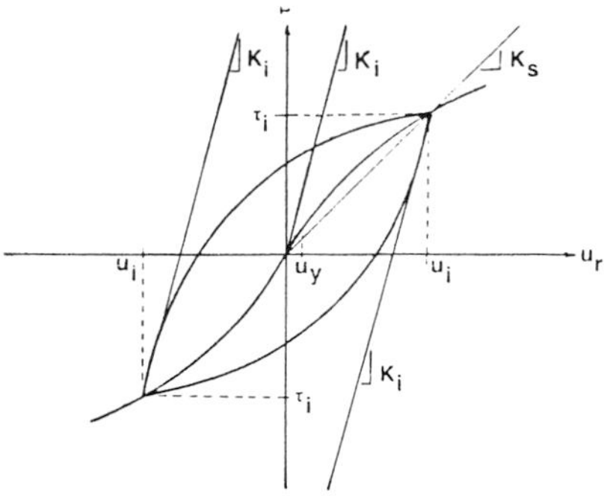

Fig. 3. Ramberg-Osgood model for cyclic stress-deformation relationship

Initial Loading or Backbone Curve:

$$u_r = u_y \left(\frac{\tau}{\delta K_i\, u_y}\right) \left(1 + \alpha \left|\frac{\tau}{K_i\, u_y}\right|^{R-1}\right) \tag{5a}$$

Unloading-Reloading Curves:

$$u_r - u_i = u_y \left(\frac{\tau - \tau_i}{\delta K_i\, u_y}\right) \left(1 + \frac{2}{2^R} \left|\frac{\tau - \tau_i}{K_i\, u_y}\right|^{R-1}\right) \tag{5b}$$

where u_r = the relative displacement of the interface, τ = the interface shear stress, and (u_i, τ_i) = the relative displacement and shear stress, respectively, at the point of loading reversal as shown in Fig. 3. The remaining terms in Eq. (5) are: K_i = initial shear stiffness; u_y = a reference displacement, and material parameters α and R. The tangent stiffness at the point of load reversal is assumed to be equal to K_i, as indicated in Fig. 3. Parameter δ is given by

$$\delta = N^t \tag{5c}$$

where N = number of cycles of loading and t is a constant that defines degradation or softening. The initial stiffness K_i is defined in terms of normal stress and given by

$$K_i = K\, \sigma_n^{\ n} \tag{5d}$$

where K and n are constants, and σ_n is the normal stress on the interface.

The shear modulus of an interface is usually smaller than that of the surrounding material (soil). If it is assumed that the slippage (failure) occurs in the soil, then as a simplification, its value may be adopted same as that of the soil. During slip defined by criteria such as Mohr-Coulomb, and with compressive normal stress, the value of the shear stiffness is reduced to a small value equal to about 0.01 x G_o, where G_o is the initial shear modulus.

Verification

The model proposed above for the shear behavior is implemented in a dynamic finite element procedure and the predictions are compared with laboratory results for (Ottawa) sand-concrete interface tested by using a cyclic shear device (ref. 3, 8, 9, 10). The interface occurs between the sand in the top half of the box and the concrete in the lower half, as shown in Fig. 4, which also shows the finite element mesh used. Tests were run with two (initial) relative densities for the sand, D_r = 15 and 80%. A normal stress equal to 28.00 psi (193 kPa) was first applied, kept constant, and then cyclic sinusoidal loading in the horizontal direction with a displacement amplitude u_r^m = 0.05 inch (1.27 mm) was applied to the top box with a frequency of 1.0 Hz. In the dynamic finite element analysis, a time step Δt = 0.05 sec was used with the Newmark β-method for time integration.

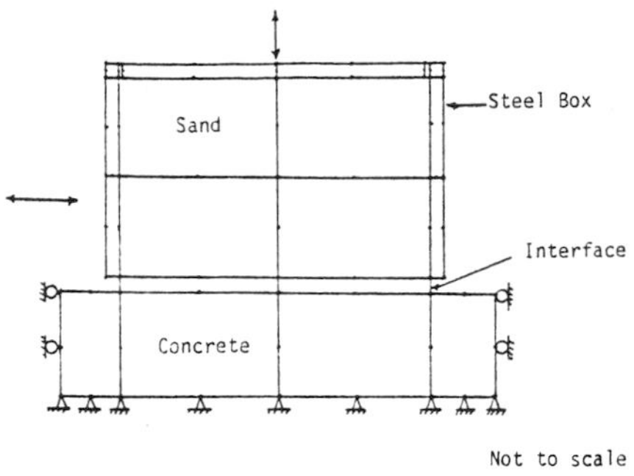

Fig. 4. Finite element mesh for simulation of cyclic test ($t'/B = 0.01$ where t' = interface thickness and B = width of element)

Material constants for the model were determined by following the procedures and reported in refs. (8, 9), and shown in Table 1 below.

TABLE 1. Parameters for model for shear behavior (ref. 8, 9)

Parameter	Relative Density	
	D_r = 15%	D_r = 80%
K	30.0	77.0
n	1.0	1.0
R	2.5	2.5
α	2.356	6.75
t	0.324	0.025
u_y	0.01 in (0.254 mm)	0.01 in (0.254 mm)

Note: All values are nondimensional, except for u_y.

Figures 5 and 6 show comparisons between predictions and observations for D_r = 15 and 80%, respectively, and for two sets of typical cycles N = 1, 2 and 10. It can be seen that the proposed model provides satisfactory simulation of the cyclic shear response of the interface.

Normal Behavior

It is relatively difficult to define the normal component $[C_n]$ for various modes of deformation. In fact, it is the normal behavior that has not been fully understood and requires careful mathematical and experimental studies. In general, the normal stiffness $[C_n]$ is dependent on various factors:

$$[C_n] = [C_n\ (v_r, \tau, \sigma_o, \gamma_i, A_c, \sigma^{ss}, N)] \quad (6)$$

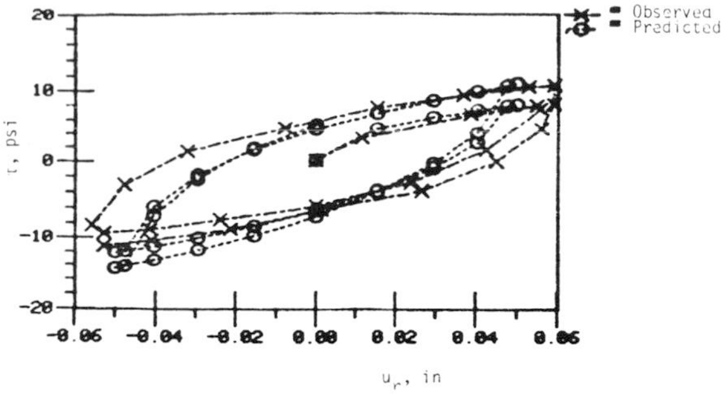

(a) Cycles 1 and 2

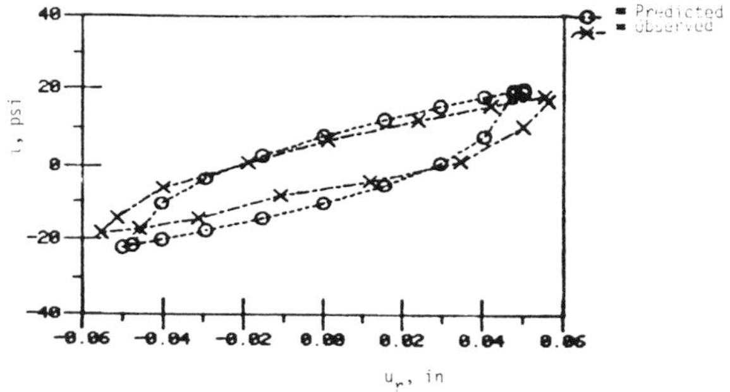

(b) Cycle 10

Fig. 5. Comparisons for shear stress vs. relative displacement responses: D_r = 15%, σ_n = 28.0 psi (1.0 inch = 25.4 mm, 1.0 psi = 6.89 kPa)

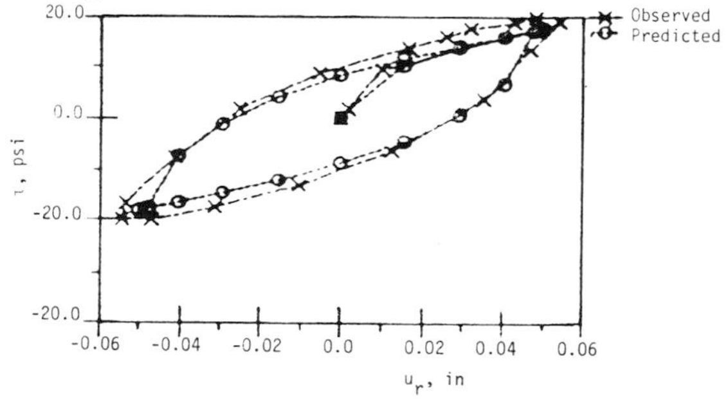

(a) Cycles 1 and 2

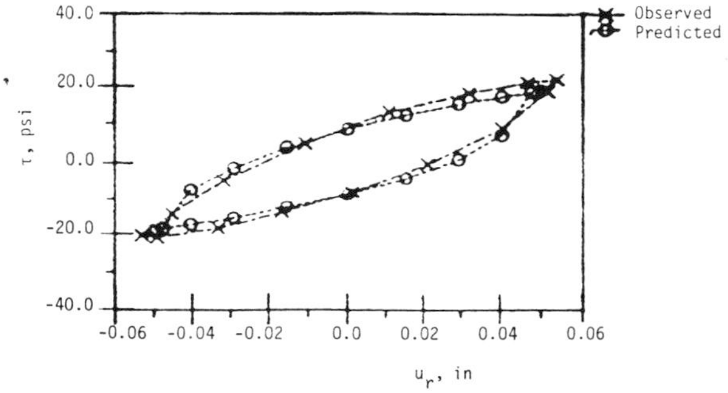

(b) Cycle 10

Fig. 6. Comparisons for shear stress vs. relative displacement responses: D_r = 80%, σ_n = 28.0 psi (1.0 inch = 25.4 mm, 1.0 psi = 6.89 kPa)

where v_r = (relative) normal displacement, σ_o = initial stress, A_c = area of (partial) contact, and σ^{ss} = stress in the surrounding elements.

Based on observed laboratory test results for static and cyclic response of the sand-concrete interface, the normal behavior is expressed in three parts, Fig. 7a, as special cases of Eq. (6):

(i) Virgin Loading

$$\sigma_n = a_o \, (e^{a_1 \varepsilon_n} - 1.0) \tag{7a}$$

where σ_n = normal stress, ε_n = normal strain, ε_m = maximum strain reached after a cycle of loading (its value is zero if no unloading has taken place), and a_o, a_1 = material constants. Here strain is defined as an (average) value equal to relative normal displacement divided by the thickness of the interface or the soil specimen. In the case of the latter, an approximation is made that the average strain is relevant to the entire interface thickness.

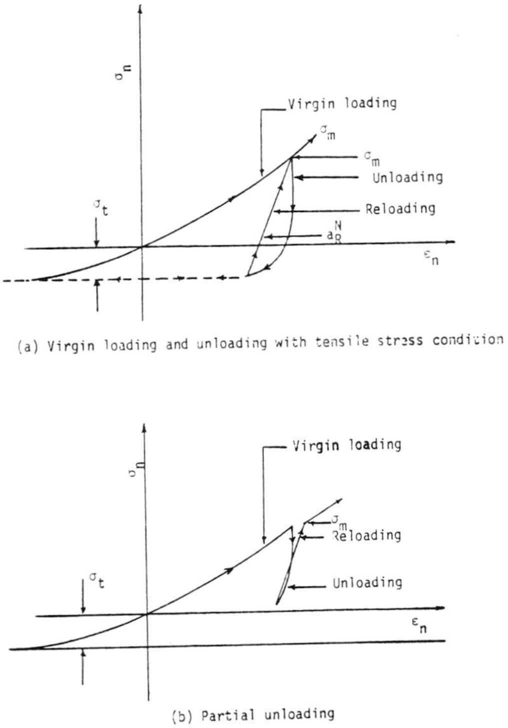

Fig. 7. Schematic of loading, unloading, reloading and tensile stress condition

(ii) Unloading

$$\sigma_n = \sigma_m - \frac{(\varepsilon_m - \varepsilon_n)}{a_2 + p_1 (\varepsilon_m - \varepsilon_n)} \quad \text{for } (\varepsilon_m - \frac{(\sigma_m + \sigma_t)}{a_R^N}) < \varepsilon_n < \varepsilon_m \tag{7b}$$

and

$$\sigma_n = -\sigma_t \quad \text{for } \varepsilon_n < (\varepsilon_m - \frac{(\sigma_m + \sigma_t)}{a_R^N}) \tag{7c}$$

where σ_m = maximum stress reached or stress at which unloading occurred, σ_t = allowable tensile strength of the interface, a_2 = material constant, and a_R^N = reloading modulus after N cycles defined below, and

$$p_1 = \frac{(1.0 - a_2 a_R^N)}{(\sigma_m + \sigma_t)} \tag{7d}$$

(iii) Reloading

$$\sigma_n = a_R (\varepsilon_n - \varepsilon_o) + \sigma_o$$

$$\text{for } (\varepsilon_m - \frac{\sigma_m + \sigma_t}{a_R^N}) < \varepsilon_n < \varepsilon_o + \frac{(\sigma_m + \sigma_t - \sigma_o)}{a_R^N} \tag{7e}$$

where a_R = Max (a_R^N, a_U), ε_o = Max $\{(\varepsilon_m - \frac{\sigma_m}{a_R^N})$, or $\varepsilon_{min}\}$, ε_{min} = minimum strain reached during unloading, σ_o = Max $\{-\sigma_t, \sigma_{min}\}$, σ_{min} = minimum stress reached during unloading, and $a_U = \frac{\partial \sigma_n}{\partial \varepsilon_n}$, Eq. (7b):

$$a_U = \frac{a_2}{[a_2 + p_1 (\varepsilon_m - \varepsilon_n)]^2} \tag{7f}$$

Effect of Loading Cycles N

Modulus a_R^N is observed to be dependent on the number of cycles and is given by

$$a_R^N = a_R^1 \, [1.0 + \frac{\bar{\delta} \, N^{a_5}}{a_3 + a_4 \, N^{a_5}}] \tag{7g}$$

where a_R^1 = initial reloading modulus, $\bar{\delta} = 0$ for static loading, and $\bar{\delta} = 1$ for for cyclic loading, and a_3, a_4, a_5 = material constants.

The initial reloading modulus is a function of initial normal stress, σ_o, and amplitude of normal cyclic load, σ_a. Investigation is in progress to evolve suitable expression for a_R^1 in terms σ_o and σ_a. In this presentation, results for only the static loading test and one cyclic loading test are presented.

Tensile Condition and Loss of Contact

The observed behavior of the sand-concrete interface shows zero tensile strength. However, to make the model more general, and to accommodate interface for different variety of materials, tensile strength denoted by σ_t is included in the model.

If, during the virgin loading, tensile conditions occur, Eq. (7a) is still valid, but with the following condition:

$$\sigma_n = - \sigma_t \quad \text{for } \varepsilon_n < \frac{\ell n \, (1.0 - \frac{\sigma_t}{a_o})}{a_1} \tag{7h}$$

Note that the previous equations for unloading and reloading already accommodate the tensile strength of the interface.

In finite element formulation, the tangent modulus for stiffness calculation is determined by using $\partial\sigma_n/\partial\varepsilon_n$ at a given load level. When $\sigma_n \geq - \sigma_t$, loss of contact is assumed to occur and the tangent modulus is reduced to a small value equal to about 0.001 x K_o, where K_o is initial modulus, Fig. 7. For a given loading, the interface element may be in bonding or debonding or partial debonding condition. To account for the partial debonding, the stiffness matrix for the interface element is evaluated only at the integration points showing contact; the latter is defined by (compressive) normal stress greater than the tensile strength of the interface.

Verification

Static Loading. Here a slow normal load is applied at the top of the sand specimen, and various cycles of unloading and reloading are performed, Fig. 8. The material constants are found from the laboratory tests and are given below:

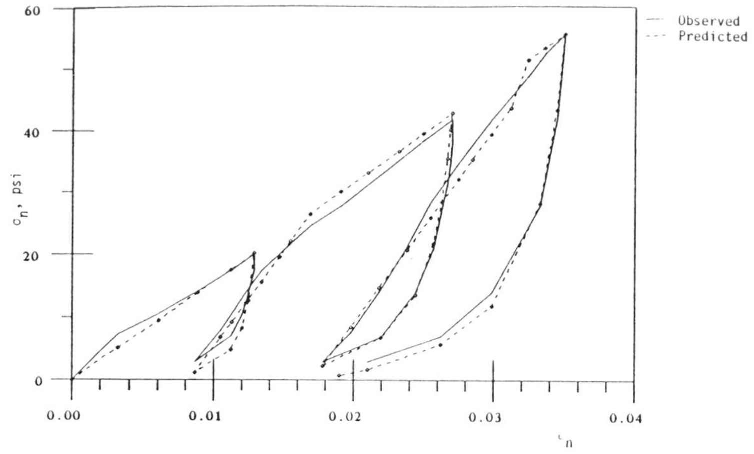

Fig. 8. Comparisons for normal stress vs. normal strain response under static loading

Virgin loading: a_o = 341475.0 psi (2345933.0 kPa) and a_1 = 0.00462,
Unloading: a_2 = 2.5636 x 10^{-5} 1/psi (3.7315 x 10^{-6} 1/kPa)
Reloading: a_R^1 = 3086.0 psi (21201.0 kPa)

Figure 8 shows a good comparison between the predictions and observations.

Cyclic Test. Various laboratory tests were performed with D_r = 15% and different values of initial σ_o and amplitudes σ_a and normal loading frequency = 1.0 Hz. A typical test with σ_o = 20 psi (138 kPa) and σ_a = 15 psi (103 kPa) is considered here. For this test, the material constants are

Virgin loading: a_o = 341475.0 psi (2345933.0 kPa) and a_1 = 0.00462,
Unloading: a_2 = 2.5636 x 10^{-5} 1/psi (3.7315 z 10^{-6} 1/kPa),
Reloading: a_R^1 = 2890 psi; a_3 = 16.73; a_4 = 3.697.0; a_5 = 2.0.

Figures 9(a) and 9(b) show comparisons between predictions and observations. The predictions are obtained by using Eqs. 7. The correlation between the two is considered satisfactory.

CONCLUSIONS

For realistic prediction of the dynamic behavior of structure-foundation systems, it is necessary to allow for nonlinear behavior of soils and interfaces. For the latter, it is required to include various deformation modes such as no-slip, slip, debonding and rebonding with appropriate shear and normal responses. When these are incorporated in solution procedures such as nonlinear finite element schemes, it is possible to obtain improved and realistic evaluation of displacements, velocities and contact pressures in dynamic loadings.

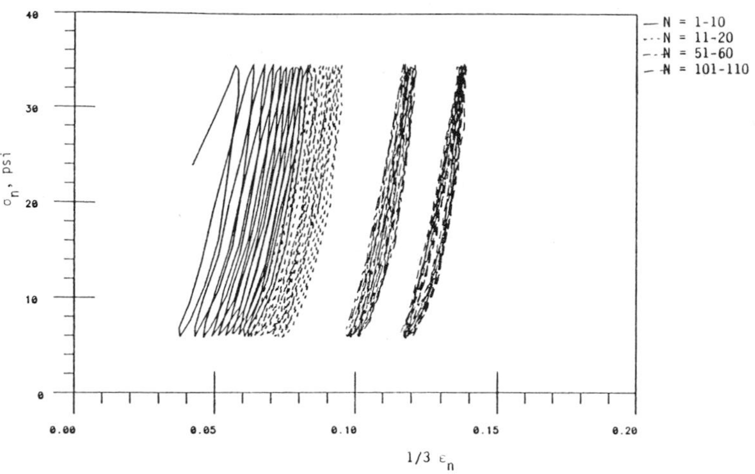

(a) Observed

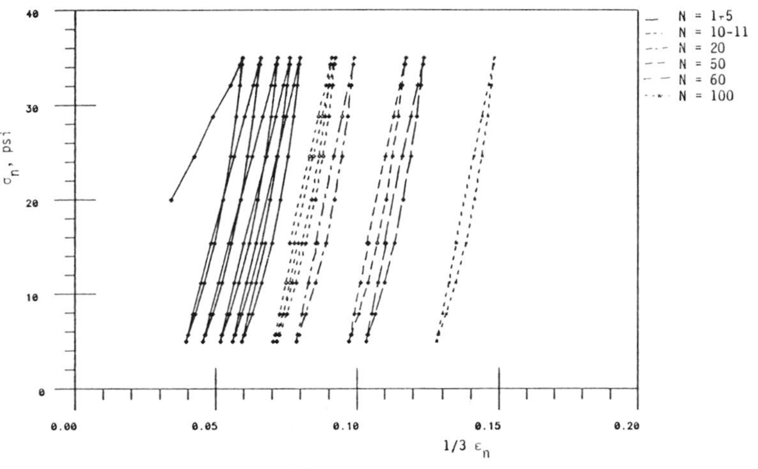

(b) Predicted

Fig. 9. Comparisons for normal stress vs. normal strain response under cyclic loading.

The interface behavior including various modes of deformation is modelled by using the thin-layer element. The constitutive description for the shear and normal responses is developed by using a piecewise linear model for loading, unloading and reloading. Material constants are determined from laboratory test results by using a new cyclic shear device for the shear and normal behavior.

The model predictions are compared with typical laboratory tests. Based on these results, it is believed that the proposed model can provide satisfactory simulation of the interface behavior. However, additional research will be needed, and is in progress for further investigation of the proposed model.

ACKNOWLEDGMENT

A part of this research was conducted under Grant No. 81-CE-94 from the National Science Foundation and under Grant No. AFOSR-83-0256 from the Air Force Office of Scientific Research, Bolling AFB. Contributions of Dr. E. C. Drumm towards the results reported in this paper are gratefully acknowledged.

REFERENCES

1 C. S. Desai, Behavior of Interfaces Between Structural and Geologic Media, Int. Conf. on Recent Advances in Geotech. Earthquake Eng. and Soil Dynamics, St. Louis, MO, 1984.
2 P. Wolf, Dynamic Soil-Structure Interaction, Prentice-Hall, Englewood Cliffs, N. J., 1985.
3 C. S. Desai, A Dynamic Multi-Degree-of-Freedom Shear Device, Report No. 80-36, Dept. of Civil Eng., VA TECH, Blacksburg, VA, 1984.
4 C. S. Desai, et al., Thin-Layer Element for Interfaces and Joints, Int. J. Num. Analyt. Methods in Geomech., 8, 1, 1984.
5 M. M. Zaman and C. S. Desai, Analysis of Interface Behavior in Dynamic Soil-Structure Interaction, 7th Int. Conf. on Struct. Mech. in Reactor Tech., Chicago, August, 1983.
6 M. M. Zaman, C. S. Desai and E. C. Drumm, Interface Model for Dynamic Soil-Structure Interaction, J. Geotech. Eng. Div., ASCE, 110, 9, 1984.
7 C. S. Desai and M. M. Zaman, Modelling and Evaluation of Interface Behavior in Dynamic Soil-Structure Interaction, Proc., 8th World Conference on Earthquake Eng., San Francisco, July, 1984.
8 C. S. Desai, E. C. Drumm and M. M. Zaman, Cyclic Testing and Modelling of Interfaces, J. Geotech. Eng. Div., ASCE, 111, 6, June, 1985.
9 E. C. Drumm and C. S. Desai, Determination of Parameters for a Model for Cyclic Behavior of Interfaces, J. of Earthquake Eng. & Struct. Dyn., accepted for publication.
10 B. K. Nagaraj, Modelling of Normal and Shear Behavior of Interface in Dynamic Soil-Structure Interaction, Ph.D. Dissertation, Dept. of Civil Engg. and Engg. Mech., Univ. of Arizona, 1985.
11 C. S. Desai, A General Basis for Yield, Failure and Potential Functions in Plasticity, Int. J. Num. Analyt. Methods in Geomech., 4, 4, 1980.
12 C. S. Desai, G. Frantziskonis and S. Somasundaram, Constitutive Modelling for Geological Materials, Proc., 5th Int. Conf. on Num. Meth. in Geomech., Nagoya, Japan, April, 1985.
13 I. M. Idriss, R. Dobry and R. D. Singh, Nonlinear Behavior of Clays During Cyclic Loading, J. Geotech. Eng. Div., ASCE, 104, 12, 1978.
14 P. C. Jennings, Periodic Response of a General Yielding Structure, J. Eng. Mech. Div., ASCE, 90, 2, 1964.

CONTACT AND INCLUSION PROBLEMS

A FINITE ELEMENT METHOD FOR CONTACT PROBLEMS WITH FINITE DEFORMATIONS AND MIXED BOUNDARY CONDITIONS

G.Z VOYIADJIS[1] and A.A. POE[1]
[1]Dept. of Civil Engineering, Louisiana State Univ., Baton Rouge, Louisiana

ABSTRACT

An incremental iterative numerical method is presented here for the solution of contact problems. A system of constitutive equations for elasto-plastic materials with finite strain is used to solve for the deformations and stresses. In the elasto-plastic analysis, the materials exhibit both kinematic and isotropic strain-hardening. The solution to a number of two-dimensional contact problems is presented here to demonstrate the accuracy and applicability of the proposed method.

INTRODUCTION

The finite element method has been extensively used in recent years for the analysis of bodies in contact subjected to various loading conditions (ref.1-19). Many of these solutions are formulated by imposing constraints or by using techniques which result in a loss of generality.

In this work, an incremental iterative finite element method of solution is presented for the elasto-plastic analysis of contact problems. A system of constitutive relations, applicable for the elasto-plastic behavior of materials, are used to obtain the deformation and stresses for the contact problems. These equations are formulated such that they can be used to solve for contact problems subjected to finite strains, and large deformations. The elasto-plastic model used exhibits both kinematic and isotropic strain hardening.

The method of solution proposed in this work, for solving contact problems, provides a different approach than the ones used in other works. The contact forces and displacements in the unknown region of contact are solved for concurrently during each iteration using a common stiffness matrix. The additional equations needed to calculate the unknown forces and displacements are provided through the compatibility of displacements and the equilibrium of contact forces in the region of contact. This analysis can be used for solving contact problems between flexible bodies subjected to large deformations.

The solution to a number of contact problems is presented to demonstrate the accuracy and applicability of the proposed method.

GEOMETRY OF CONTACT REGION

The geometry of contact described in this work is for the region of contact between two bodies in contact. This procedure could be generalized such that a body can have several regions of contact with different bodies. It is also assumed that the two bodies will initiate contact between them before the procedure outlined below is followed. This initial contact may be limited to a single point. The geometry in this work is confined to two-dimensional problems.

The region of contact between bodies A and B is shown in Fig. 1. The body forced into contact is referred to as body A. The nodes in body A are m and those of body B are n-m (where $n > m$). The finite element nodes in body A are expressed as A_i, where i varies from 1 to m, and those in body B are B_i, where i varies from m+1 to n.

In the case of the nodes of body A that come into contact with body B, the subscript i in A_i varies from a to b. The subscript a stands for the first node in body A that will come into contact with body B. The subscript b stands for the last possible node in body A that may come into contact with body B during the entire history of loading of this contact problem. During any given iteration, the contact nodes of body A are expressed as A_i where i varies from a to c (where $c \leq b$) as shown in Fig. 2. Intermediate nodes between A_a and A_c may not be in contact during any given iteration of a load increment. Nodes A_a through A_b should be numbered consecutively in order to minimize the bandwidth of the global stiffness matrix of bodies A and B.

During any given iteration k, it is essential to know the coordinates of the contact nodes in bodies A and B from the two previous iterations k-1 and k-2. Those from the k-1 iteration are referred to as the current coordinates of node i and are expressed as x_i and y_i. In the case of the k-2 iteration, the coordinates are referred to as the previous coordinates of node i and are expressed as x_i^p and y_i^p.

The nodes on the contact side of an eight-node isoparametric element of body B are designated as L_i, O_i and K_i, where i denotes the node from body A in contact with that side of the finite element. This is shown in Fig. 2.

Shown in Fig. 3(a) are the lines joining the previous and current coordinates of nodes A_i and A_{i+1}. These lines intersect two different sides of the contact elements on body B. The three nodes on each side of the contact element of body B are designated as L_i, O_i, K_i and L_{i+1}, O_{i+1}, K_{i+1}. The subscript i denotes the contact node A_i whose previous and current coordinates form the line crossing the side of the contact element in B. The special case when the lines joining the previous and current coordinates of two different nodes A_i and A_{i+1} intersect the same side of the contact

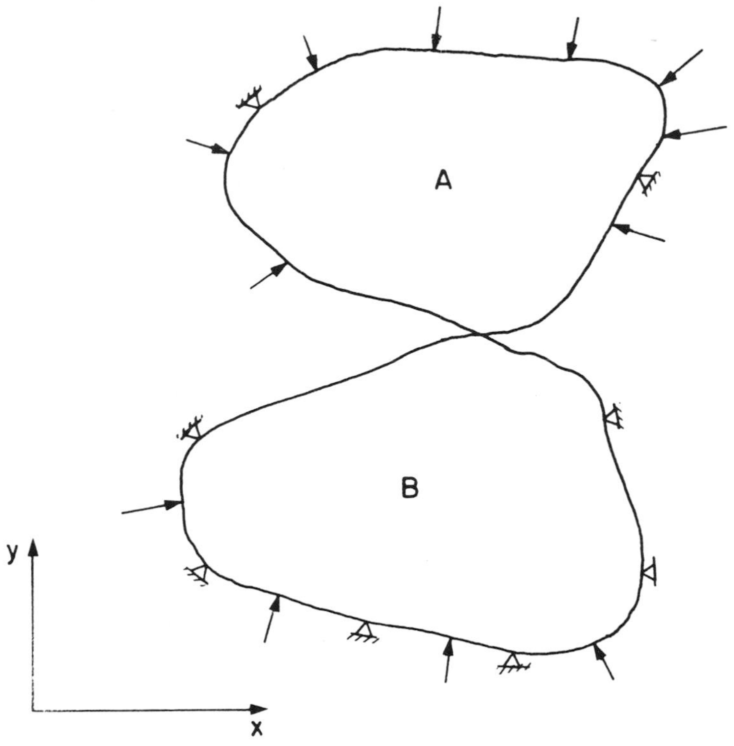

Fig. 1. Body A forced into contact with body B.

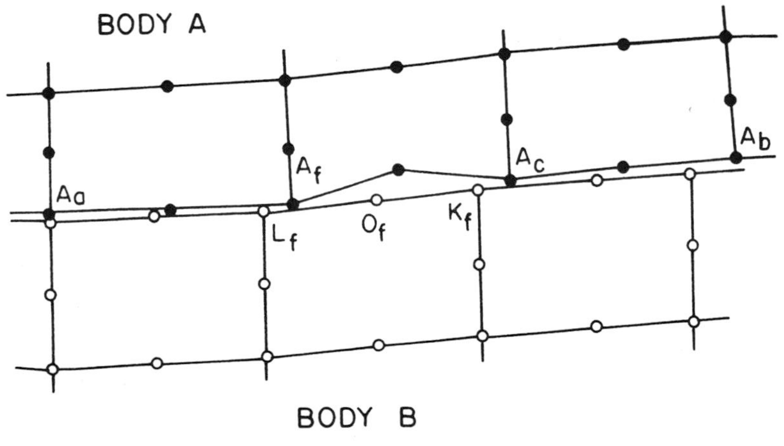

Fig. 2. Contact node numbering scheme.

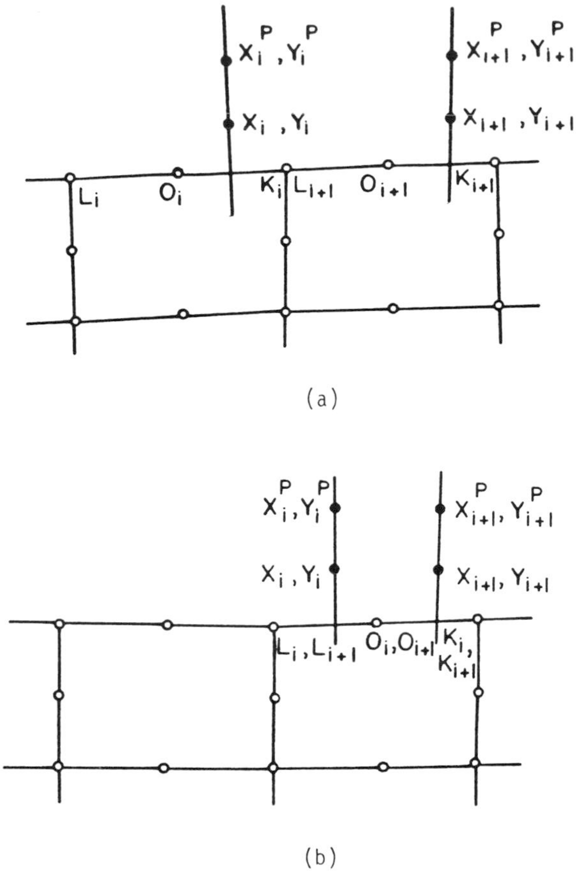

Fig. 3. Contact node numbering scheme for body B.

element in B is shown in Fig. 3(b). The side of the contact element in B will be referred to as the contact side element.

The current coordinates x and y of the nodes in the contact region of body B, namely L_i, O_i and K_i, are expressed as $x(L_i)$, $x(O_i)$, $x(K_i)$ and $y(L_i)$, $y(O_i)$, $y(K_i)$, respectively.

The equation of the line joining the previous and current coordinates of node A_i is written as

$$y - y_i^p = m_i(x - x_i^p) \tag{1}$$

where m_i is the slope of the line, given by

$$m_i = \frac{y_i - {}^p y_i}{x_i - x_i^p} \tag{2}$$

The equation given by (1) can be rewritten in the slope-intercept form as

$$y = m_i\, x + c_i \tag{3}$$

The quantity denoted by c_i is found from (1);

$$c_i = y_i^p - m_i\, x_i^p \tag{4}$$

Both c_i and m_i are constant for the line joining the previous and current coordinates of a specific contact node A_i in one particular iteration. The constants c_i and m_i are different for each different contact node A_i and are recomputed during each iteration. This is also true for the previous and current coordinates x_i^p, y_i^p and x_i, y_i, respectively.

The lines forming each contact side-element are second-order curves. This is true because 8-node finite elements are used in this work. The algorithm can easily be modified to accommodate any type element which is used in the contact region of body B. One approximation describing the displacements of three points lying on a second-order curve is

$$y = \alpha_f + \beta_f\, x + \gamma_f\, x^2 \tag{5}$$

The terms α_f, β_f and γ_f are constant for a specific contact side-element during a given iteration. These terms are calculated for each contact side-element during each iteration. The values of α_f, β_f and γ_f are found from the current coordinates of the nodes L_f, 0_f and K_f. The current y coordinates of each node are known and are written in matrix form as

$$\begin{Bmatrix} y(L_f) \\ y(0_f) \\ y(K_f) \end{Bmatrix} = \begin{bmatrix} 1 & x(L_f) & [x(L_f)]^2 \\ 1 & x(0_f) & [x(0_f)]^2 \\ 1 & x(K_f) & [x(K_f)]^2 \end{bmatrix} \begin{Bmatrix} \alpha_f \\ \beta_f \\ \gamma_f \end{Bmatrix} \tag{6}$$

The three unknown quantities in (6) are α_f, β_f and γ_f. These are calculated using the matrix equations

$$\begin{Bmatrix} \alpha_f \\ \beta_f \\ \gamma_f \end{Bmatrix} = \begin{bmatrix} 1 & x(L_f) & [x(L_f)]^2 \\ 1 & x(0_f) & [x(0_f)]^2 \\ 1 & x(K_f) & [x(K_f)]^2 \end{bmatrix}^{-1} \begin{Bmatrix} y(L_f) \\ y(0_f) \\ y(K_f) \end{Bmatrix} \tag{7}$$

The point of intersection of the line joining the previous and current coordinates of each contact node A_i and the contact side-element which this line intersects is found by solving (3) and (5) simultaneously for x and y. The solution yields two values for x and y because the resulting equation is a quadratic equation found to be

$$x_{i1}^*,\ x_{i2}^* = \frac{-(\beta_f - m_i) \pm [(\beta_f - m_i^2) - 4\gamma_f(\alpha_f - c_i)]^{1/2}}{2\gamma_f} \tag{8}$$

Corresponding values of y_{i1}^* and y_{i2}^* are found by substituting x_{i1}^* and x_{i2}^* into (3) or (5).

The equation of the line given by (5) is for a line which is infinitely long; the contact side-element is only a small segment of this line. This means that the line joining the previous and current coordinates of a contact node A_i has two points in common with each line whose equation is given by (5). The only possible exception is examined later.

In Fig. 4(a), the line joining the previous and current coordinates of a contact node A_i is shown intersecting a line on which a contact side-element lies in two different points. Neither of these points of intersection is on the contact side-element. In Fig. 4(b), one of the points of intersection is on the contact side-element. To determine which contact side-element the line joining the previous and current coordinates of contact node A_i will intersect, the inequalities

$$x(L_f) \leq x_{i1}^* \leq x(K_f) \tag{9a}$$

and

$$x(L_f) \leq x_{i2}^* \leq x(K_f) \tag{9b}$$

are checked for all contact side-elements. The situation shown in Fig. 4(a) is the case if neither (9a) nor (9b) are satisfied. If either (9a) or (9b) are satisfied, then the particular contact side-element which is intersected by the line joining the previous and current coordinates of a node A_i is determined. This is the case shown in Fig. 4(b).

The contact side-element which is intersected by the line joining the previous and current coordinates of a contact node A_i is found for all nodes A_i that are not in contact during the current iteration. For each contact node A_i, one point of intersection is found for the line joining the previous and current coordinates of A_i and the contact side-element that this line intersects. The x coordinate of the point of intersection is x_{i1}^* or x_{i2}^* as given by (9a) or (9b). The correct x and y coordinates of the point of intersection are denoted as x_i^* and y_i^* where y_i^* is found by substituting x_i^* into (3)

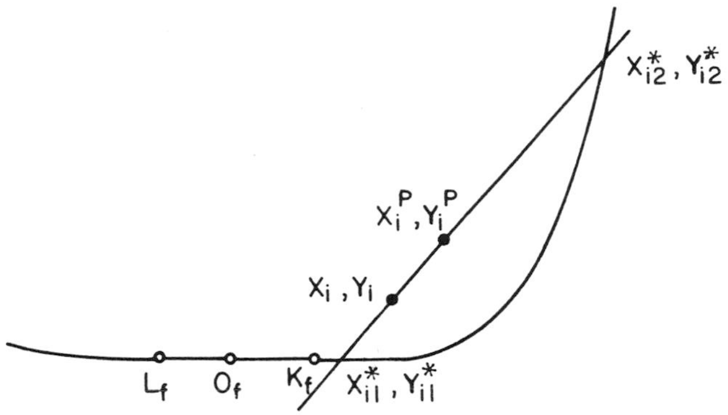

(a) Line joining previous and current coordinates of contact node A_i intersecting two points which do not lie on the contact side-element.

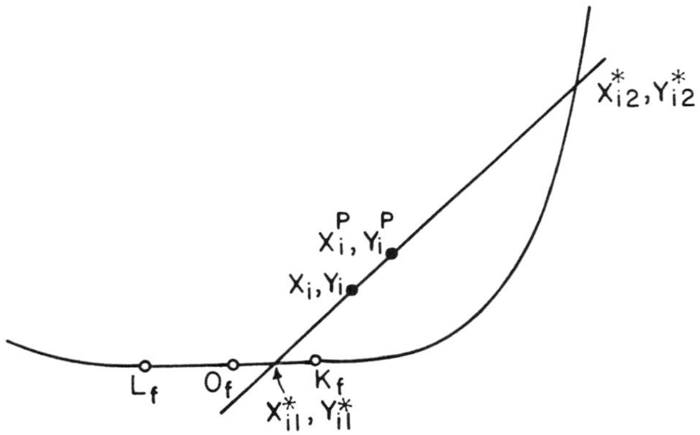

(b) Line joining previous and current coordinates of contact node A_i intersecting two points, one of which is on the contact side-element.

Fig. 4. Identifying the correct solution to equation (8).

or (5). The coordinates x^*_{i1} and y^*_{i1} shown in Fig. 4(b) are also the desired coordinates x^*_i and y^*_i. Once the contact side-element which is intersected by the line joining the previous and current coordinates of node A_i is determined, the three nodes lying on the contact side-element are recorded as L_i, O_i and K_i, where i corresponds to the contact node A_i.

The second step determines if any nodes from the contact region of body A come into contact with body B. This step makes use of the results obtained in the first step.

Shown in Fig. 5(a) are the previous and current positions of a contact node A_i. The line joining these two positions is shown intersecting a contact side-element. The distance between the previous and current positions of the contact node A_i is

$$D_i^1 = [(x_i - x_i^p)^2 + (y_i - y_i^p)^2]^{1/2} \tag{10}$$

Since the loading is applied in small increments, the straight line joining the previous and current positions will be nearly the same as the actual path of the node. The length of the line joining the previous position of the node and the point of intersection which has coordinates (x_i^*, y_i^*) is

$$D_i^2 = [(x_i^* - x_i^p)^2 + (y_i^* - y_i^p)^2]^{1/2} \tag{11}$$

The contact node A_i comes into contact with body B if

$$D_i^1 \geq D_i^2 \tag{12}$$

If equation (12) is not true, then the contact node A_i does not come into contact with body B. This condition corresponds to the situation shown in Fig. 5(a). If it is determinated that a node does not come into contact during an iteration, the algorithm simply treats that node like all other nodes which are not in contact. Each node which may possibly come into contact with body B (nodes A_a to A_b) is checked for contact during each iteration of every load increment.

If equation (12) is true, a contact condition as shown in Fig. 5(b) exists. The node A_i is constrained to lie on the contact side-element and has coordinates x_i^*, y_i^*. The lengths shown in Fig. 5(b) denoted as S_i and ηS_i must be determined in order to calculate the shape functions.

To determine S_i and ηS_i, the differential lengths are integrated over the entire length. The differential length of a curve is given by

$$d\ell = [(dx)^2 + (dy)^2]^{1/2} \tag{13}$$

This can be rewritten as

$$d\ell = [1 + (\frac{dy}{dx})^2]^{1/2} \, dx \tag{14}$$

For a second-order curve such as the curve given by equation (5), the term $(dy/dx)^2$ is given by

$$(\frac{dy}{dx})^2 = 4\,\gamma^2\,x^2 + 4\,\gamma\,\beta\,x + \beta^2 \tag{15}$$

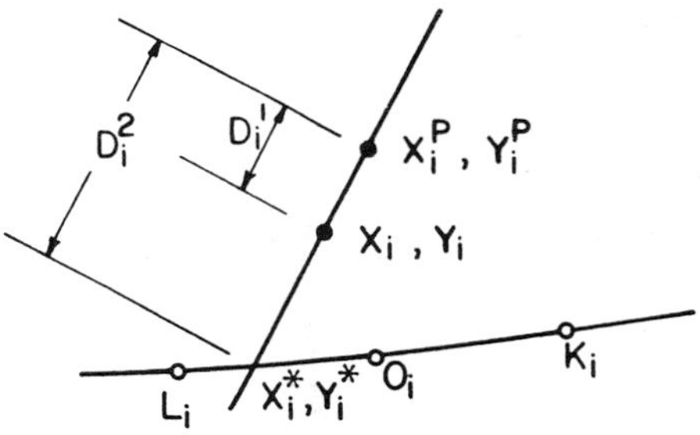

(a) Node A_i not in contact with B.

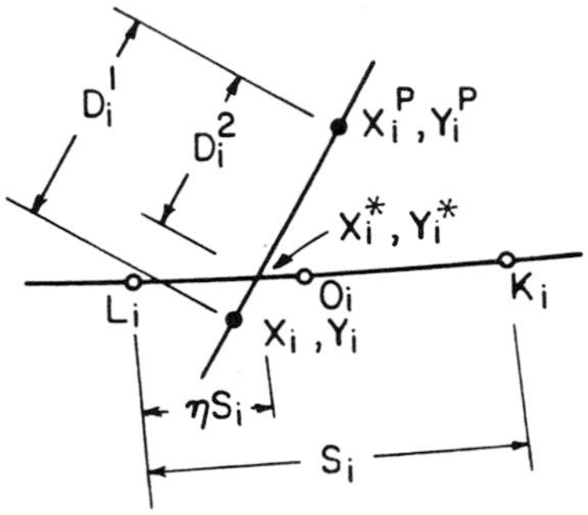

(b) Node A_i in contact with B.

Fig. 5. Determining if contact occurs.

This leads to the general expression for the length S_i of a second-order curve evaluated between arbitrary points P and Q

$$S_{PQ} = \int_{x_P}^{x_Q} [4\gamma^2 x^2 + 4\gamma\beta x + (1+\beta^2)]^{1/2} dx \tag{16}$$

Using equation (16), the lengths S_i and ηS_i can be found;

$$S_i = \int_{x(L_i)}^{x(K_i)} [4\gamma^2 x^2 + 4\gamma\beta x + (1+\beta^2)]^{1/2} dx \tag{17a}$$

and

$$\eta S_i = \int_{x(L_i)}^{x_i^*} [4\gamma^2 x^2 + 4\gamma\beta x + (1+\beta^2)]^{1/2} dx \tag{17b}$$

These lengths are found for each node A_i which is in contact with any contact side-element. The lengths S_i and ηS_i are all measured from the node L_i of the contact side-element under consideration.

Using (17a) and (17b), the three shape functions for each contact node A_i are

$$N1_i = 1 - 3\eta + 2\eta^2 \tag{18a}$$

$$N2_i = 4\eta - 4\eta^2 \tag{18b}$$

$$N3_i = 2\eta^2 - \eta \tag{18c}$$

These shape functions are used to relate the displacements and contact forces of contact nodes A_i to the displacements and contact forces of the nodes on the contact line-element with which the contact nodes A_i are in contact.

The contact forces on nodes L_i, O_i and K_i resulting from the contact force on node A_i are found from the shape functions of the element as shown in Fig. 6(a). The compatibility expression of the displacement of node A_i and the three nodes of the contact side-element is shown in Fig. 6(b). These relationships are utilized in the development of the finite element formulation.

The procedure for determining if a contact node A_i comes into contact with body B includes calculating the slope of the line joining the previous and current positions of node A_i as given previously. The slope of this line is given by expression (2). If the line joining the previous and current coordinates of a contact node A_i is a vertical line, then (2) cannot be used. For a case such as this, x_i^* is found to be

$$x_i^* = x_i = x_i^p \tag{19}$$

The coordinate y_i^* can be found once x_i^* is known by solving (5) for y_i^*;

$$y_i^* = \alpha_i + \beta_i\, x_i^* + \gamma_i (x_i^*)^2 \tag{20}$$

CONSTITUTIVE RELATIONS

The constitutive model used in this work is for an elasto-plastic material with kinematic and isotropic strain hardening. The Lagrangian reference frame is used in this formulation. The constitutive relations are described in terms of the material strain e_{AB}, and the second Piola-Kirchhoff stress tensor s_{AB}. An objective material stress rate tensor $\dot{s}_{AB}$ is used, where

$$\dot{s}_{AB} = \frac{\partial s_{AB}}{\partial t} \tag{21}$$

The resulting expression between $\dot{s}_{AB}$ and $\dot{e}_{AB}$ is given as (ref.20-22)

$$\dot{s}_{AB} = D_{ABCD}\ \dot{e}_{AB} \tag{22}$$

where D is the elasto-plastic stiffness matrix which corresponds to the yield function $f(s_{AB}, A_{AB}, \kappa, C_{AB}, J)$ given in references (20-22). D is expressed as

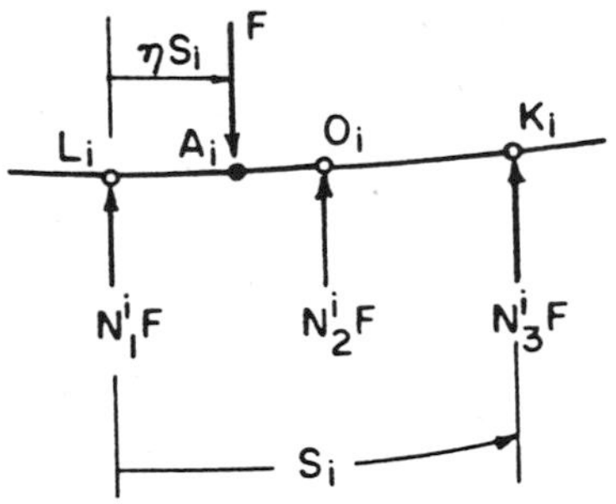

(a) Use of shape functions to determine contact forces on nodes L_i, O_i and K_i resulting from contact force on node A_i.

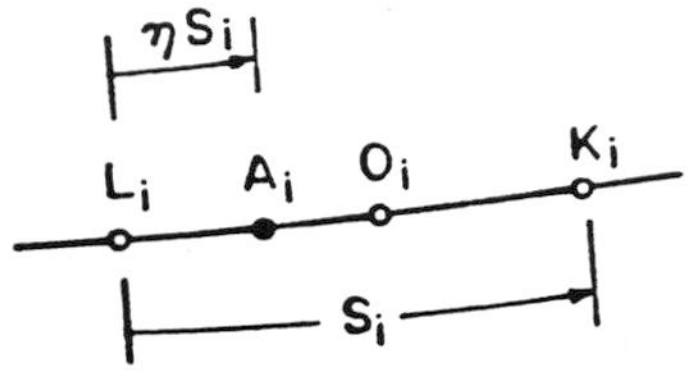

$$\Delta A_i = N_1^i \, \Delta L + N_2^i \, \Delta O + N_3^i \, \Delta K$$

(b) Use of shape functions to determine the displacement of node A_i in terms of the displacement of nodes L_i, O_i and K_i.

Fig. 6. Contact forces and displacements.

$$D_{MNPQ} = E_{MNPQ} - E_{MNCD} \left[\frac{\frac{\partial f}{\partial s_{AB}} \frac{\partial f}{\partial s_{CD}} E_{ABPQ} + \frac{\partial f}{\partial e_{PQ}} \frac{\partial f}{\partial s_{CD}} + \frac{\partial f}{\partial s_{CD}} \frac{\partial f}{\partial J} R_{PQ}}{Q} \right] \tag{23}$$

where

$$R_{PQ} = [2\delta_{PQ} + 4\delta_{PQ} \, e_{KK} - 4e_{PQ} + 8e_{QR} \, e_{RP} - 8e_{PQ} \, e_{KK}$$

$$- 4\delta_{PQ} \, e_{OS} \, e_{OS} + 4\delta_{PQ} \, e_{LL} \, e_{KK}]/2J \tag{24}$$

and

$$Q = E_{ABCD} \frac{\partial f}{\partial s_{CD}} \frac{\partial f}{\partial s_{AB}} - \frac{\partial f}{\partial \kappa} s_{AB} \frac{\partial f}{\partial s_{AB}} J^{-1}$$

$$- \frac{\partial f}{\partial A_{AB}} (s_{AB} - A_{AB}) b \frac{\frac{\partial f}{\partial s_{MN}} \frac{\partial f}{\partial s_{MN}}}{(s_{QR} - A_{QR}) \frac{\partial f}{\partial s_{QR}}} \qquad (25)$$

This formulation gives the correct interpretation of the material behavior in the Eulerian reference frame. It also by-passes the use of the Jaumann stress rate which is shown to be inaccurate when used in conjunction with kinematic hardening plasticity at finite strains.

FINITE ELEMENT IMPLEMENTATION

The eight-node isoparametric element is used in the formulation of this work. Only the basic aspects in the finite element formulation related to this work will be dealt with in this paper. For details, the reader is referred to numerous papers and monographs (e.g., ref.23).

The governing finite element equations for bodies A and B are expressed as follows:

$$[K_A] \{\Delta q_A\} = \{\Delta R_A\} + \{\Delta F_A\} \qquad (26a)$$

and

$$[K_B] \{\Delta q_B\} = \{\Delta R_B\} + \{\Delta F_B\} \qquad (26b)$$

where $[K_A]$, $[K_B]$, $\{\Delta q_A\}$, $\{\Delta q_B\}$, $\{\Delta F_A\}$, $\{\Delta F_B\}$, $\{\Delta R_A\}$, $\{\Delta R_B\}$ are the tangent stiffness matrices, incremental nodal displacement vectors, incremental nodal point contact force vectors, and nodal point external forces for bodies A and B, respectively. In this work the formulation is used to solve for two-dimensional problems only.

Equations (26a) and (26b) are combined and expressed as

$$\begin{bmatrix} K_A & 0 \\ 0 & K_B \end{bmatrix} \begin{Bmatrix} \Delta q_A \\ \Delta q_B \end{Bmatrix} - \begin{Bmatrix} \Delta F_A \\ \Delta F_B \end{Bmatrix} = \begin{Bmatrix} \Delta R_A \\ \Delta R_B \end{Bmatrix} \qquad (27)$$

with the total number of equations in expression (27) equal to twice the number of nodes in bodies A and B. The total number of unknowns in (27) is equal to twice the number of nodes in bodies A and B plus twice the number of contact forces for the nodes in contact between bodies A and B.

Using the relationships shown in Fig. 6(a) between the nodal point contact forces which act on body B to the nodal point contact forces which act on A, expression (27) may be rewritten as

$$\begin{bmatrix} K_A & 0 \\ 0 & K_B \end{bmatrix} \begin{Bmatrix} \Delta q_A \\ \Delta q_B \end{Bmatrix} - [D] \begin{Bmatrix} \Delta F_A \\ 0 \end{Bmatrix} = \begin{Bmatrix} \Delta R_A \\ \Delta R_B \end{Bmatrix} \qquad (28)$$

All terms in [D] are zero except the following:

$$[D]_{2A_i,2A_i} = [D]_{2A_i-1,2A_i-1} = 1 \qquad \text{for } i = a \text{ to } c \qquad (29a)$$

$$[D]_{2L_i,2A_i} = [D]_{2L_i-1,2A_i-1} = -\,N1_i \qquad \text{for } i = a \text{ to } c \qquad (29b)$$

$$[D]_{2O_i,2A_i} = [D]_{2O_i-1,2A_i-1} = -\,N2_i \qquad \text{for } i = a \text{ to } c \qquad (29c)$$

$$[D]_{2K_i,2A_i} = [D]_{2K_i-1,2A_i-1} = -\,N3_i \qquad \text{for } i = a \text{ to } c \qquad (29d)$$

Similarly using the relationship shown in Fig. 6(b), the nodal displacements in the x and y direction for each contact node in body A can be expressed in terms of the displacements of the contact nodes in body B. Therefore, equation (28) may be expressed as

$$\begin{bmatrix} K_A & 0 \\ 0 & K_B \end{bmatrix} [\bar{D}] \begin{Bmatrix} \Delta q_R \\ \Delta q_B \end{Bmatrix} - [D] \begin{Bmatrix} \Delta F_A \\ 0 \end{Bmatrix} = \begin{Bmatrix} \Delta R_A \\ \Delta R_B \end{Bmatrix} \qquad (30)$$

where $\{\Delta q_R\}$ is the reduced incremental nodal displacement vector. All terms in $[\bar{D}]$ are zero except the following:

$$[\bar{D}]_{2A_i,2A_j} = [\bar{D}]_{2A_i-1,2A_j-1} = 1 \qquad \text{for } i = 1 \text{ to } m,\; j = 1 \text{ to } m,\; j \neq a \text{ to } c \qquad (31a)$$

$$[\bar{D}]_{2A_i,2L_j} = [\bar{D}]_{2A_i-1,2L_j-1} = N1_i \qquad \text{for } i = a \text{ to } c,\; j = a \text{ to } c \qquad (31b)$$

$$[\bar{D}]_{2A_i,2O_j} = [\bar{D}]_{2A_i-1,2O_j-1} = N2_i \qquad \text{for } i = a \text{ to } c,\; j = a \text{ to } c \qquad (31c)$$

$$[\bar{D}]_{2A_i,2K_j} = [\bar{D}]_{2A_i-1,2K_j-1} = N3_i \qquad \text{for } i = a \text{ to } c,\; j = a \text{ to } c \qquad (31d)$$

$$[\bar{D}]_{2A_i,2A_j} = [\bar{D}]_{2A_i-1,2A_j-1} = 1 \qquad \text{for } i = (m+1) \text{ to } n,\; j = (m+1) - c \text{ to } n \qquad (31e)$$

Expression (30) may be further reduced by proper matrix multiplication to the following:

$$[\bar{K}_{AB}] \left\{\frac{\Delta q_R,\ \Delta F_A}{\Delta q_B}\right\} = \left\{\frac{\Delta R_A}{\Delta R_B}\right\} \qquad (32)$$

This system of equations may now be solved for the unknown nodal displacements in bodies A and B less the nodal contact displacements of body A. In addition, these equations are also solved concurrently for the nodal contact forces in body A. The displacements of the contact nodes on body A and the contact forces on the contact nodes of body B are found by using the relationships shown in Fig. 6(a) and 6(b), respectively.

SPHERE ON A SEMI-INFINITE BODY

The discretized model of this problem is shown in Fig. 7. The axisymmetric, elastic solution of this problem using the finite element procedure outlined in this work is compared to Hertz's closed form solution. Table 1 shows the applied load P and the corresponding radius of contact for both the finite element solution and Hertz's solution. It also tabulates the corresponding deflection of the initial point of contact. From this table, we can see that the radius of contact obtained from the numerical solution varies from 94 to 99 percent of the corresponding radius contact obtained from the Hertz solution. The deflection of the initial point of contact compares well with the Hertz solution up to $a/r < 0.32$.

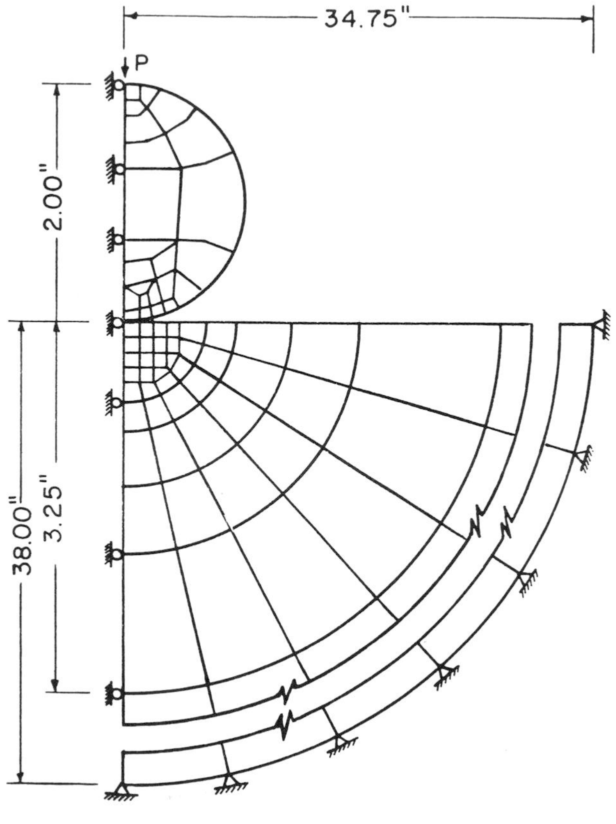

Fig. 7. Finite element mesh for sphere on half-space.

TABLE 1

Values for the deflection of the initial point of contact and the radius of contact for increasing loads.

Load (lb)	v_o (in)	a_H (in)	a (in)
2,700	0.0026	0.0618	0.0650
22,000	0.0114	0.1243	0.1306
70,000	0.0243	0.1828	0.1941
190,000	0.0464	0.2550	0.2574
380,000	0.0737	0.3212	0.3191
650,000	0.1050	0.3842	0.3807

v_o ≡ deflection of initial point of contact.
a_H ≡ radius of contact due to Hertz's solution.
a ≡ radius of contact due to numerical solution.

TABLE 2

Distribution of vertical contact nodal forces for elastic, small-strain solution of steel sphere ($E = 30 \times 10^6$ psi) forced into aluminum half-space ($E = 10 \times 10^6$ psi).

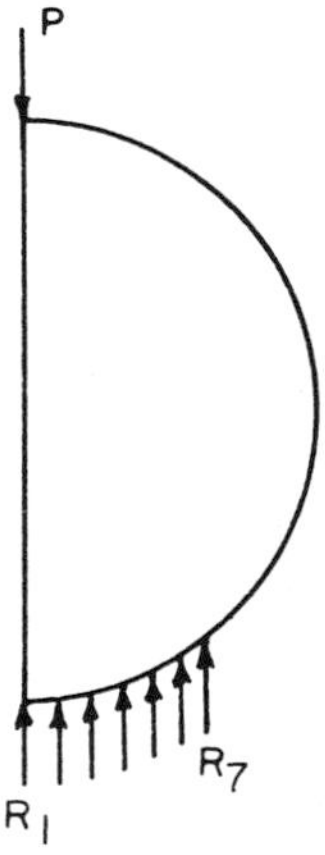

All R_i are shown after being multiplied by $2\pi r$.

Load (kips)	2.5	2.7	22	70	190	380	650
R_1	2.5	2.689	4.57	6.578	9.382	11.830	14.286
R_2		0.011	17.395	28.112	42.185	55.780	68.100
R_3			0.035	35.287	50.617	68.970	85.950
R_4				0.023	87.769	129.182	174.418
R_5					0.497	114.206	141.475
R_6						0.032	165.688
R_7							0.083

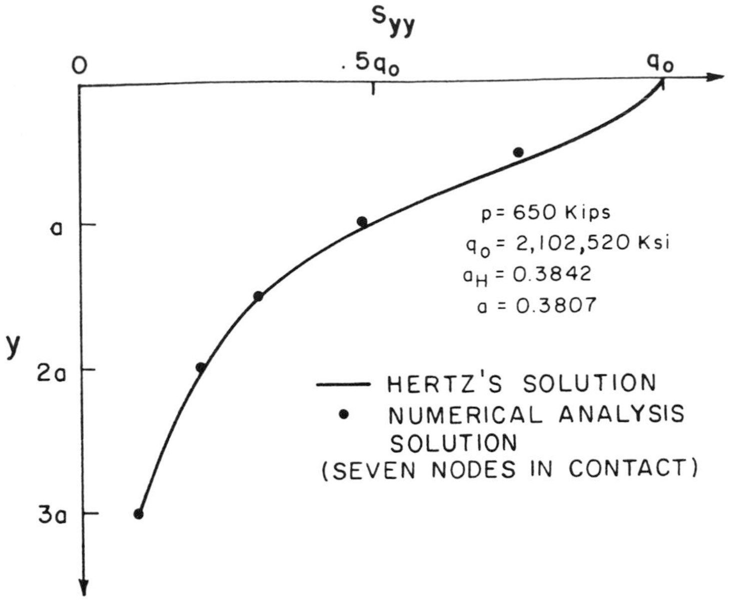

Fig. 8. Results from elastic analysis.

Table 2 shows the equilibrating contact nodal forces from the numerical solution of the problem.

Figure 8 compares the normal stress in the y-direction between the finite element solution and Hertz's solution for an applied load of P = 650,000 lb. The stresses are plotted for the semi-infinite body along the axis of symmetry. The maximum pressure obtained from Hertz's solution at the surface of contact is 2.1×10^6 ksi. At this load, seven nodes are in contact. The results indicate that this numerical procedure accurately predicts the stresses. The material is not allowed to yield even though it has reached such high stress levels. This is done in an effort to compare this procedure with a credible solution such as the Hertz solution and at the same time allowing a large number of nodes to be in contact.

SPHERE ON PLATE

The finite element model of this problem is shown in Fig. 9. The analysis for this problem is limited to two nodes in contact. The distribution of the normal stress in the y-direction in the plate is shown in Fig. 10. No further analysis is conducted for this problem due to the lack of a closed form solution to compare with the numerical results.

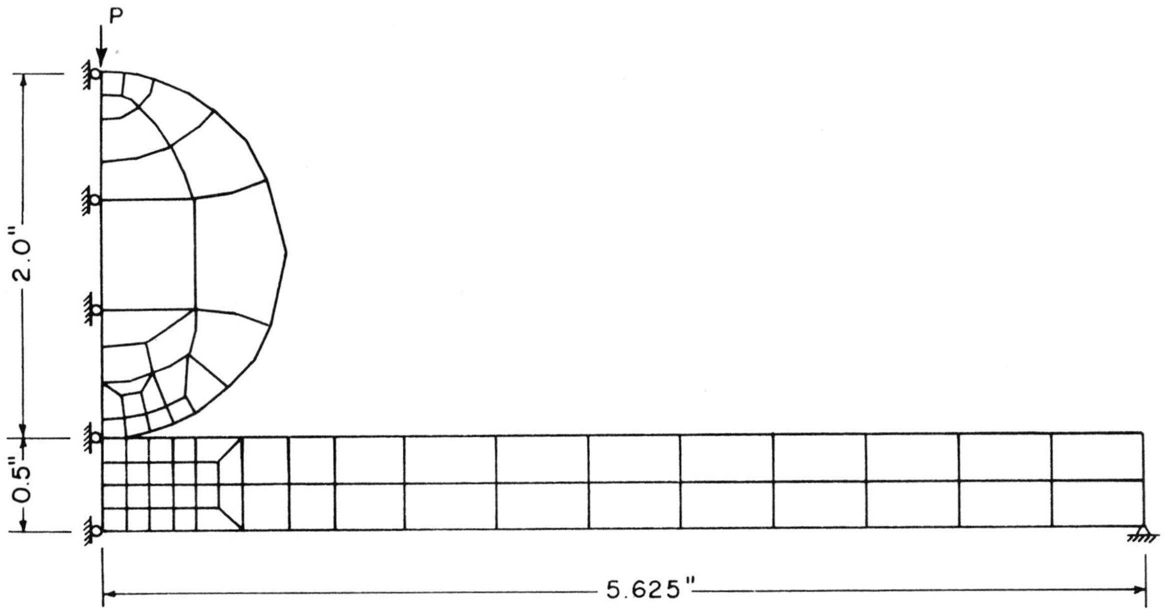

Fig. 9. Finite element mesh for sphere on plate.

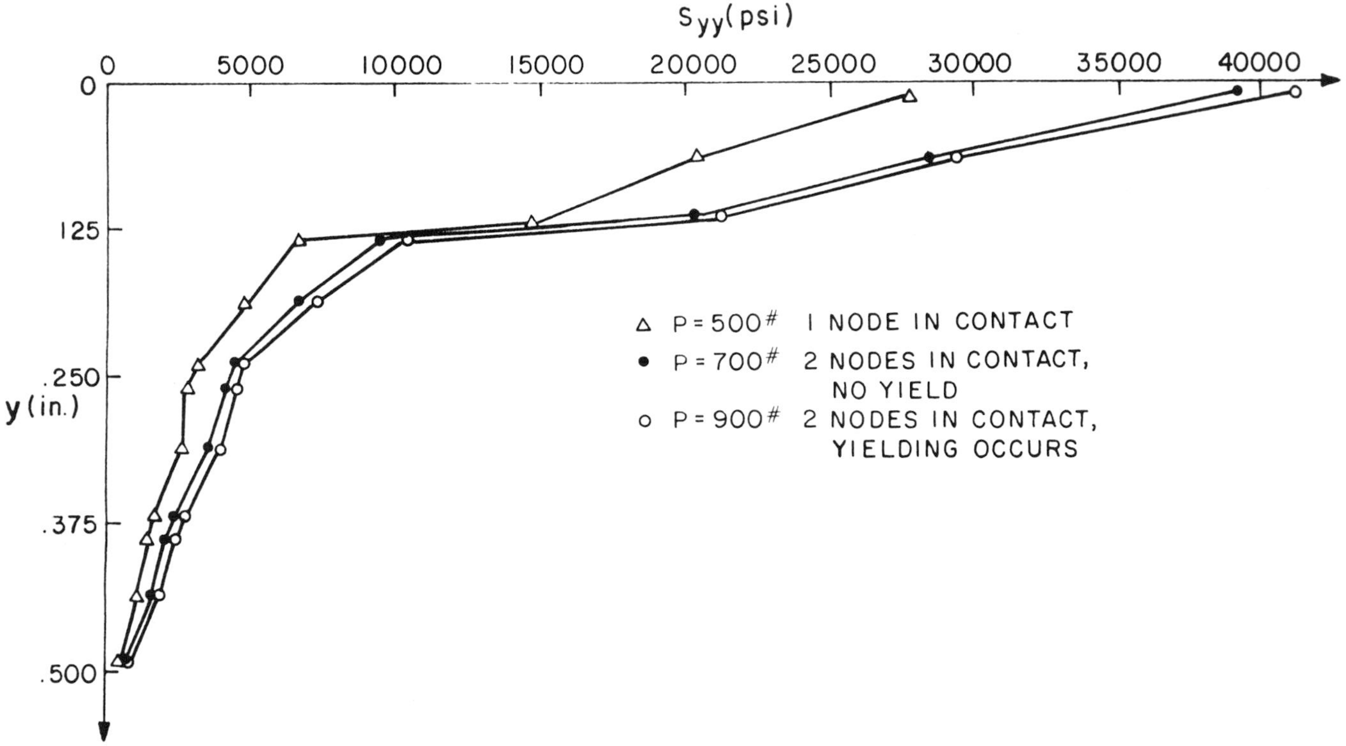

Fig. 10. Results from elasto-plastic analysis.

CONCLUSION

An incremental iterative numerical procedure is outlined in this work for the solution of contact problems. The contact forces and displacements in the unknown region of contact are solved concurrently during each iteration using a common stiffness matrix. The additional equations required to calculate the unknown forces and displacements are provided through the compatibility of displacements and the equilibrium of contact forces in the region of contact.

The applicability of this approach to two-dimensional problems is demonstrated through the solution of two contact problems.

A more general and accurate formulation for the geometry of contact used in this work would be to replace the suggested second order curve with that obtained from the element shape functions. Due care should be given in this case to insure that the shape functions are defined in the Lagrangian reference frame and then properly transformed to the current reference frame. If the shape functions are immediately used in the current reference frame, then that will lead to erroneous results unless the mid-node on the side of the element is always located at mid-distance between the corner nodes.

REFERENCES

1 E.A. Wilson and B. Parsons, Finite Element Analysis of Elastic Contact Problems Using Differential Displacement, International Journal for Numerical Methods in Engineering, 2 (1970) 387-395.
2 S.K. Chan and I.S. Tuba, I. S., A Finite Element Method for Contact Problems of Solid Bodies - Part I. Theory and Validation, International Journal of Mechanics and Science, 13 (1971) 615-625.
3 C. Hardy, C.N. Baronet and G.V. Tordion, The Elasto-Plastic Indentation of a Half-Space by a Rigid Sphere, International Journal for Numerical Methods in Engineering, 3 (1971) 451-462.
4 S. Ohte, Finite Element Analysis of Elastic Contact Problems, Bull. Japanese Society of Mechanical Engineers, 16 (1973) 797-804.
5 A. Francavilla and O.C. Zienkiewicz, A Note on Numerical Computation of Elastic Contact Problems, International Journal for Numerical Methods in Engineering, 9 (1975) 913-924.
6 J.L. Urzua and O.A. Pecknold, Analysis of Frictional Contact Problems using an Interface Element, Proceedings of the Symposium on Applications of Computer Methods in Engineering, August 23-26, Los Angels, CA, 1977.
7 T.J.R. Hughes, R.L. Raylor and W. Kanoknukulchai, A Finite Element Method for Large Displacement Contact and Impact Problems, in: Formulations and Computational Algorithms in Finite Element Analysis (K. J. Bathe, et al., Eds.), M.I.T. Press, 1977.
8 L.R. Herrmann, Finite Element Analysis of Contact Problems, Journal of the Engineering Mechanics Division, ASCE, 104 (1978) 1043-1057.
9 N. Okamoto and M. Nakazawa, Finite Element Incremental Contact Analysis with Various Frictional Conditions, International Journal for Numerical Methods in Engineering, 14 (1979) 337-359.
10 N. Kikuchi and J.T. Oden, Contact Problems in Elasticity - A Study of Variational Inequalities and Finite Element Methods for a Class of Contact Problems in Elasticity, TICOM Report 79-8, 1979.

11 J.J. Kalker, The Computation of Three-Dimensional Rolling Contact with Dry Friction, International Journal of Numerical Methods in Engineering, 14 (1979) 1293-1307.
12 H.S. Cheng and L.M. Keer (Eds.), Solid Contact and Lubrication, AMD, 39, American Society of Mechanics Engineers, 1980.
13 J. Tseng and M.D. Olson, The Mixed Finite Element Method Applied to Two-Dimensional Elastic Contact Problems, International Journal for Numerical Methods in Engineering, 17 (1981) 991-1014.
14 T.H.H. Pian and K. Kubomura, Formulation of Contact Problems by Assumed Stress Hybrid Elements, in: Nonlinear Finite Element Analysis in Structural Mechanics, (W. Wunderlich, et al., Eds.), Springer-Verlag, 1981.
15 L.T. Campos, J.T. Oden and N. Kikuchi, A Numerical Analysis of a Class of Contact Problems with Friction in Elastostatics, Computer Methods in Applied Mechanics Engineering, 34 (1982) 821-845.
16 G.Z. Voyiadjis and N.E. Buckner, Indentation of a Half-Space with a Rigid Indentor, International Journal for Numerical Methods in Engineering, 19 (1983) 1555-1578.
17 G.Z. Voyiadjis and S. Navaee, Finite Strain Contact Problem of Cylinder Embedded in Body, Journal of the Engineering Mechanics Division, ASCE, 110 EM11 (1984) 1597-1609.
18 K.J. Bathe and A. Chaudhary, A Solution Method for Planar and Axisymmetric Contact Problems, International Journal for Numerical Methods in Engineering, 21 (1985) 65-88.
19 C.S. Desai, M.M. Zaman, J.G. Lightner and H.J. Siriwardane, Thin Element for Interfaces and Joints, International Journal of Analytical and Numerical Methods in Geomechanics (in press).
20 G.Z. Voyiadjis, Experimental Determination of the Material Parameters of Elasto-Plastic, Work-Hardening Metal Alloys, Materials Science and Engineering Journal, 62 (1984) 99-107.
21 G.Z. Voyiadjis and P.D. Kiousis, Stress Rate and the Lagrangian Formulation of the Finite-Strain Plasticity for a Von Mises Kinematic Hardening Model, submitted for publication to the International Journal of Solids and Structures.
22 P.D. Kiousis, G.Z. Voyiadjis and M.T. Tumay, A Large Strain Theory for Two Dimensional Problems in Geomechanics, International Journal of Analytical and Numerical Methods in Geomechanics (in press).
23 O.C. Zienkiewicz, The Finite Element Method in Engineering Science, McGraw-Hill, New York, 1977.

EFFECT OF INTERFACE PROPERTIES ON WAVE PROPAGATION IN A MEDIUM WITH INCLUSIONS

S.K. DATTA[1] and H.M. LEDBETTER[2]

[1]Department of Mechanical Engineering and Cooperative Institute for Research in Environmental Sciences, University of Colorado, Boulder, Colorado 80309 (USA)

[2]Fracture and Deformation Division, National Bureau of Standards, Boulder, Colorado 80303 (USA)

ABSTRACT

This study considers propagation of effective plane longitudinal and shear waves through a medium with a random distribution of spherical inclusions. We assume that inclusions and matrix possess different elastic properties and that a thin layer of elastic material with still different properties separates the inclusions from the matrix. Also, we assume same-size inclusions and same-thickness layers. We find that the layers substantially affect the phase velocities and attenuation of coherent plane waves propagating through this composite medium.

INTRODUCTION

Much current practical interest exists concerning wave propagation through a composite medium with a random distribution of inclusions: particles, flakes, or long continuous fibers in a homogenous matrix. Several theoretical studies report wave speeds and attenuation of coherent plane elastic waves propagating through an elastic homogeneous medium containing spherical or ellipsoidal elastic inclusions (refs. 1-20). All these studies assume that the inclusions bond perfectly with the surrounding matrix material. Recently, Sayers (ref. 21) examined the effect of a distribution of thin elastic spherical shells in a homogeneous matrix material of different elastic properties.

In the present study, we analyze a problem different from the one considered in (ref. 21). We assume that the inclusions possess elastic properties different from those of the matrix. Also, we consider the effect of a thin boundary layer of material through which the elastic properties vary either continuously or discontinuously from those of the inclusions to those of the matrix. Results for the case without a thin layer are obtained by letting the layer thickness go to zero.

The object of this study is to explore the practicality of using ultrasound to characterize properties of interface layers in a particulate composite. Ultrasound is a practical tool for measuring properties of, and characterizing the state of, a material with microstructure (or changes in microstructure). References to these various studies occur in (ref. 21). In the following section, we derive the equations governing scattering by an inclusion surrounded by a thin layer of variable elastic properties.

SCATTERING BY A SPHERICAL INCLUSION SURROUNDED BY A THIN LAYER

Consider a spherical inclusion of radius a and elastic properties, λ_1, μ_1, ρ_1 embedded in an elastic matrix of material properties λ_2, μ_2, ρ_2. Also, let the inclusion be separated from the matrix by a thin layer of uniform thickness $h(\ll a)$ but variable material properties $\lambda(r)$, $\mu(r)$ and $\rho(r)$. Here, λ, μ denote Lamé constants and ρ the density. Let $\lambda(r)$, $\mu(r)$ be expressed as

$$\lambda(r) + 2\mu(r) = (\lambda_1' + 2\mu_1')\, f(r), \quad a < r < a + h \tag{1}$$

$$\mu(r) = \mu_1'\, g(r), \quad a < r < a + h \tag{2}$$

f(r) and g(r) are general functions of r. A special case arises when

$$f(a) = \frac{\lambda_1 + 2\mu_1}{\lambda_1' + 2\mu_1'}, \quad g(a) = \frac{\mu_1}{\mu_1'}$$

$$f(a + h) = \frac{\lambda_2 + 2\mu_2}{\lambda_1' + 2\mu_1'}, \quad g(a + h) = \frac{\mu_2}{\mu_1'} \tag{3}$$

with the stipulation that f(r) and g(r) with their first derivatives are continuous in (a, a + h). Since h is assumed to be much smaller than a, it follows from (3) that f'(a) and g'(a) can be approximated by

$$f'(a) = \frac{(\lambda_2 + 2\mu_2) - (\lambda_1 + 2\mu_1)}{h(\lambda_1' + 2\mu_1')}$$

$$g'(a) = \frac{\mu_2 - \mu_1}{h\mu_1'} \tag{4}$$

Note that λ_1' and μ_1' are Lamé constants of the interface material at some value of r ($a < r < a + h$), say at $r = a + h/2$.

Another special case would be that the interface material possesses constant properties. Then we have

$$f(r) = g(r) = 1, \quad a < r < a + h \tag{5}$$

We also make the assumption that h is very much smaller than the wavelength of the propagating wave. Then, to first order in h/λ, λ being the wavelength,

$$\tau^t_{rr} = \tau^s_{rr} + \tau^i_{rr}, \quad \tau^t_{r\theta} = \tau^s_{r\theta} + \tau^i_{r\theta}$$

$$\tau^t_{r\phi} = \tau^s_{r\phi} + \tau^i_{r\phi} \tag{6}$$

Here τ_{ij} is the stress tensor and superscripts t, s, and i denote the transmitted, scattered, and incident field quantities, respectively. Note that τ^s_{ij}, τ^i_{ij}, and τ^t_{ij} appearing above are calculated at $r = a$. The spherical polar coordinates r, θ, ϕ are defined in Fig. 1. Boundary conditions (6) express the fact that, to first order in h/λ, the traction components do not suffer any jump across the layer. However, the displacement components suffer jumps given by

$$u^s_r + u^i_r - u^t_r = \frac{hK_1}{\lambda_1' + 2\mu_1'}\tau^t_{rr} \tag{7}$$

$$u^s_\theta + u^i_\theta - u^t_\theta = \frac{hK_2}{\mu_1'}\tau^t_{r\theta} \tag{8}$$

$$u^s_\phi + u^i_\phi - u^t_\phi = \frac{hK_2}{\mu_1'}\tau^t_{r\phi} \tag{9}$$

where

$$K_1 = \int_0^1 \frac{dx}{f(a + hx)}, \quad K_2 = \int_0^1 \frac{dx}{g(a + hx)} \tag{10}$$

Using equations (3) and (4) in (10), it can be shown that K_1 and K_2 are given approximately by

$$K_1 = \frac{\lambda_1' + 2\mu_1'}{\lambda_2 + 2\mu_2 - (\lambda_1 + 2\mu_1)} \ln\left(1 + \frac{\lambda_2 + 2\mu_2 - (\lambda_1 + 2\mu_1)}{\lambda_1 + 2\mu_1}\right) \tag{11}$$

$$K_2 = \frac{\mu_1'}{\mu_2 - \mu_1} \ln\left(1 + \frac{\mu_2 - \mu_1}{\mu_1}\right) \tag{12}$$

On the other hand, if eq. (5) is used then

$$K_1 = K_2 = 1 \tag{13}$$

Mal and Bose (ref. 5) studied a problem similar to the one considered here. They assumed a thin viscous third layer between the sphere and the matrix and imposed the condition of continuity of radial displacement.

The incident wave will be assumed to be either a plane longitudinal wave propagating in the positive z-direction or a plane shear wave polarized in the x-direction and propagating in the positive z-direction. Thus,

$$\underset{\sim}{u}^i = e^{ik_1 z}\underset{\sim}{e}_z + e^{ik_2 z}\underset{\sim}{e}_x \tag{14}$$

where $k_1 = \omega/c_1$ and $k_2 = \omega/c_2$. ω denotes the circular frequency of the wave and c_1, c_2 denote the longitudinal and shear wave speeds in the matrix. The factor $e^{-i\omega t}$ has been suppressed.

$\underset{\sim}{u}^i$ given above can be expanded in spherical vector wave functions as

$$\underset{\sim}{u}^i = \frac{1}{ik_1} \sum_{n=0}^{\infty} i^n (2n+1) \underset{\sim}{L}^{(1)}_{on} +$$

$$\frac{1}{2i} \sum_{n=1}^{\infty} \sum_{m=-1}^{\infty} \frac{2n+1}{n(n+1)} i^n \left[\underset{\sim}{M}^{(1)}_{mn} (\delta_{m1} + n(n+1)\,\delta_{m,-1})\right.$$

$$\left. + \frac{1}{k_2} \underset{\sim}{N}^{(1)}_{mn} (\delta_{m1} - n(n+1)\,\delta_{m,-1})\right] \tag{15}$$

Vector wave functions $\underset{\sim}{L}^{(1)}$, $\underset{\sim}{M}^{(1)}$ and $\underset{\sim}{N}^{(1)}$ appearing above are regular at $r = 0$ and are given by

$$\underset{\sim}{L}^{(1)}_{mn} = [\underset{\sim}{e}_r \frac{\partial}{\partial r} j_n(k_1 r) P_n^m(\cos\theta) + \underset{\sim}{e}_\theta j_n(k_1 r) \frac{1}{r}\frac{\partial}{\partial\theta} P_n^m(\cos\theta)$$

$$+ \underset{\sim}{e}_\phi \frac{im}{r\sin\theta} j_n(k_1 r) P_n^m(\cos\theta)]\, e^{im\phi}$$

$$\underset{\sim}{M}_{mn} = [\underset{\sim}{e}_\theta \frac{im}{\sin\theta} j_n(k_2 r) P_n^m(\cos\theta) - \underset{\sim}{e}_\phi j_n(k_2 r) \frac{\partial}{\partial\theta} P_n^m(\cos\theta)]\, e^{im\phi}$$

$$\underset{\sim}{N}^{(1)}_{mn} = [\underset{\sim}{e}_r \frac{n(n+1)}{r} j_n(k_2 r) P_n^m(\cos\theta) + \underset{\sim}{e}_\theta \frac{1}{r}\frac{\partial}{\partial r}(r j_n(k_2 r)) \times$$

$$\frac{\partial}{\partial\theta} P_n^m(\cos\theta) + \underset{\sim}{e}_\phi \frac{im}{r\sin\theta}\frac{\partial}{\partial r}(r j_n(k_2 r)) P_n^m(\cos\theta)]\, e^{im\phi} \qquad (16)$$

Thus, the scattered and transmitted fields can be written

$$\underset{\sim}{u}^s = \sum_{n=0}^{\infty} \sum_{m=-1}^{1} [A_{mn} \underset{\sim}{L}^{(3)}_{mn} \delta_{mo} + B_{mn} \underset{\sim}{M}^{(3)}_{mn} + C_{mn} \underset{\sim}{N}^{(3)}_{mn}] \qquad (17)$$

$$\underset{\sim}{u}^t = \sum_{n=0}^{\infty} \sum_{m=-1}^{1} [A'_{mn} \underset{\sim}{L}^{(1)\prime}_{mn} \delta_{mo} + B'_{mn} \underset{\sim}{M}^{(1)\prime}_{mn} + C'_{mn} \underset{\sim}{N}^{(1)\prime}_{mn}] \qquad (18)$$

where the prime denotes that k_1 and k_2 are to be replaced by k_1' ($= \omega/c_1'$) and k_2' ($= \omega/c_2'$), respectively. c_1' and c_2' are the wave speeds in the inclusion. $\underset{\sim}{L}^{(3)}$, $\underset{\sim}{M}^{(3)}$ and $\underset{\sim}{N}^{(3)}$ are obtained by replacing j_n by h_n in equation (16). Note that j_n is the spherical Bessel function of the first kind and h_n is the spherical Hankel function of the first kind.

The constants A, B, C, A', B', C' are found by the use of the conditions (7) and (8)-(10). For this purpose, we define the following matrices:

$$M_n = \begin{bmatrix} F_n & G_n \\ H_n & I_n \end{bmatrix} \qquad (19)$$

$$L_n = \begin{bmatrix} SF_n & SG_n \\ SH_n & SI_n \end{bmatrix} \qquad (20)$$

where

$$F_n(k_1a) = nh_n(k_1a) - k_1ah_{n+1}(k_1a)$$

$$G_n(k_2a) = n(n+1)\ h_n(k_2a),\quad H_n(k_1a) = h_n(k_2a)$$

$$I_n(k_2a) = (n+1)\ h_n(k_2a) - k_2ah_{n+1}(k_2a)$$

$$SF_n(k_1a) = (n^2 - n - \tfrac{1}{2}k_2^2a^2)\ h_n(k_1a) + 2k_1ah_{n+1}(k_1a)$$

$$SG_n(k_2a) = n(n+1)\ [(n-1)\ h_n(k_2a) - k_2ah_{n+1}(k_2a)]$$

$$SH_n(k_1a) = (n-1)\ h_n(k_1a) - k_1ah_{n+1}(k_1a)$$

$$SI_n(k_2a) = (n^2 - 1 - \tfrac{1}{2}k_2^2a^2)\ h_n(k_2a) + k_2ah_{n+1}(k_2a)$$

Equations for the determination of A_{mn} and B_{mn} are found to be

$$\left[\frac{2h}{a}\kappa L_n - M_n + \frac{\mu_2}{\mu_1}M_n' L_n'^{-1} L_n\right]\begin{Bmatrix}A_{mn}\\C_{mn}\end{Bmatrix} =$$

$$a\begin{Bmatrix}u^{i}_{r(mn)}\\u^{1i}_{\theta(mn)}\end{Bmatrix} - \frac{a^2}{2\mu_2}\left[\frac{2h}{a}\kappa + \frac{\mu_2}{\mu_1}M_n' L_n'^{-1}\right]\begin{Bmatrix}\tau^{i}_{rr(mn)}\\\tau^{1i}_{r\theta(mn)}\end{Bmatrix} \tag{21}$$

Here

$$\kappa = \begin{bmatrix}\dfrac{\mu_2}{\lambda_1' + 2\mu_1'}K_1 & 0\\ 0 & \dfrac{\mu_2}{\mu_1'}K_2\end{bmatrix}$$

and M_n', L_n' are obtained from M_n, L_n, respectively, by replacing h_n and h_{n+1} by j_n and j_{n+1}, respectively, in (19) and (20), and by replacing k_1 and k_2 by k_1' and k_2', respectively. In writing (21) we express u_r^i and u_θ^i given by (15) as

$$u_r^i = \sum_{n=0}^{\infty}\sum_{m=-1}^{1} u^i_{r(mn)}\,P_n^m(\cos\theta)e^{im\phi}$$

$$u_\theta^i = \sum_{n=0}^{\infty}\sum_{m=-1}^{1}\left\{u^{1i}_{\theta(mn)}\frac{\partial P_n^m}{\partial\theta} + u^{2i}_{\theta(mn)}\frac{im}{\sin\theta}P_n^m\right\}e^{im\phi} \tag{22}$$

It then follows that

$$\begin{Bmatrix} \tau^{i}_{rr(mn)} \\ \tau^{1i}_{\theta(mn)} \end{Bmatrix} = \frac{2\mu_2}{a} \tilde{L}_n \tilde{M}^{-1}_n \begin{Bmatrix} u^{i}_{r(mn)} \\ u^{i}_{\theta(mn)} \end{Bmatrix} \tag{23}$$

$\tilde{L}_n$ and $\tilde{M}_n$ are obtained from L_n and M_n, respectively, by replacing h_n and h_{n+1} by j_n and j_{n+1}, respectively.

The equation to find B_{mn} is

$$\Bigg[\frac{h}{a} \kappa_{22} \left((n - 1) h_n(k_2a) - k_2ah_{n+1}(k_2a)\right) - h_n(k_2a) + \frac{\mu_2}{\mu_1} j_n(k_2'a) \frac{(n - 1) h_n(k_2a) - k_2ah_{n+1}(k_2a)}{(n - 1) j_n(k_2'a) - k_2'aj_{n+1}(k_2'a)} \Bigg] B_{mn} \tag{24}$$

$$= u^{2i}_{\theta(mn)} - \frac{a}{\mu_2} \left[\frac{h}{a} \kappa_{22} + \frac{\mu_2}{\mu_1} \frac{j_n(k_2'a)}{(n - 1) j_n(k_2'a) - k_2'a\, j_{n+1}(k_2'a)} \right] \tau^{2i}_{r\theta(mn)}$$

Here

$$\tau^{2i}_{r\theta(mn)} = \frac{\mu_2}{a} \frac{(n - 1) j_n(k_2a) - k_2aj_{n+1}(k_2a)}{j_n(k_2a)} u^{2i}_{\theta(mn)} \tag{25}$$

Once A_{mn}, C_{mn}, and B_{mn} are determined by solving equations (21) and (24), the scattered field is then found from equation (17). Since the expressions for the field inside the inclusion will not be needed for the derivation of the dispersion equation governing the effective wave number of plane-wave propagation through the composite medium, we do not give these here. In the following we derive equations governing propagation of effective plane waves through a medium composed of a random homogeneous distribution of identical spherical inclusions surrounded by the layers as discussed above.

PROPAGATION OF EFFECTIVE PLANE WAVES THROUGH THE COMPOSITE MEDIUM

To derive approximately the phase velocities of plane waves moving through the composite medium, we assume that wavelengths are long compared to the radius of each inclusion. In this long-wavelength limit it can be shown that, correct to $O(\varepsilon^3)$,

$$A_{mn} = i\varepsilon^3[P_n\Phi_{mn} + Q_nX_{mn}] \tag{26a}$$

$$C_{mn} = i\tau^3\varepsilon^3[R_n\Phi_{mn} + S_nX_{mn}] \tag{26b}$$

where

$$P_o = \frac{1}{3}\frac{3\lambda_2 + 2\mu_2 - (3\lambda_1 + 2\mu_1)}{4\mu_2 + (3\lambda_1 + 2\mu_1)} \frac{\left\{1 - \frac{h}{a}\frac{3\lambda_2 + 2\mu_2}{\lambda_1' + 2\mu_1'} K_1\right\}}{\left\{1 + 4\frac{h}{a}\frac{\mu_2}{\lambda_1' + 2\mu_1'} K_1\right\}}$$

$$Q_o = R_o = S_o = 0$$

$$P_1 = \frac{1}{2\tau} Q_1 = \tau R_1 = S_1 = \frac{1}{9}(\rho_1/\rho_2 - 1)$$

$$P_2 = \frac{1}{3\tau^2} Q_2 = 2R_2 = \frac{2}{3\tau^2} S_2 = \frac{2}{3}(1 - 2\sigma_2)\frac{A}{B}$$

$$A = 1 - \mu_1/\mu_2 + 2\left\{\frac{1}{\alpha}(\mu_1/\mu_2 - \sigma_1) + \frac{h}{a}\frac{\mu_1}{\mu_1'}\left(\frac{K_2}{\alpha} + \frac{\mu_1'}{\lambda_1' + 2\mu_1'}K_1\right)\right\}$$

$$B = \frac{\mu_1}{\mu_2}(8 - 10\sigma_2) + 7 - 5\sigma_2 - \frac{1}{\alpha}\left\{\frac{\mu_1}{\mu_2}(7 - 11\sigma_2) - 2\sigma_1(5\sigma_2 - 7)\right\}$$
$$- \frac{2h}{a}\frac{\mu_1}{\mu_1'}\left(\frac{K_2}{\alpha} + \frac{\mu_1'}{\lambda_1' + 2\mu_1'}K_1\right)(5\sigma_2 - 7)$$

$$\frac{1}{\alpha} = \frac{12\frac{h}{a}\frac{\mu_1}{\mu_1'}\left(K_2 - \frac{\mu_1'}{\lambda_1' + 2\mu_1'}K_1\right)}{2(7 - 10\sigma_1) + \frac{\mu_1}{\mu_2}(5 + \sigma_1) + 12\frac{h}{a}\frac{\mu_1}{\mu_1'}\left(\frac{1}{3}K_2(7 + 2\sigma_1) + \frac{\mu_1'}{\lambda_1' + 2\mu_1'}K_1\right)}$$

$$\tau = k_2/k_1, \quad \varepsilon = k_1 a$$

$$\Phi_{mn} = (2n + 1)\, i^n \frac{1}{ik_1}\delta_{mo}$$

$$X_{mn} = \frac{2n + 1}{n(n + 1)}\, i^n \frac{1}{2ik_2}\left\{\delta_{m1} - n(n + 1)\,\delta_{m,-1}\right\}$$

In writing the expressions for A_{mn} and C_{mn} we followed the notation of Mal and Bose (ref. 5) for easy comparison with their results. They assumed $K_1 = 0$ and $K_2 \neq 0$.

Once the scattered field due to a single inclusion is known, multiple scattering due to a number of inclusions can easily be calculated. In particular, following the steps discussed before by us (refs. 8 and 9) it can be shown that effective speeds of propagation of plane longitudinal and shear waves are given by

$$\frac{k_1^{*2}}{k^2} = \frac{(1 + 9cP_1)(1 + 3cP_o)\left\{1 + \frac{3c}{2} P_2 (2 + 3\tau^2)\right\}}{1 - 15cP_2 (1 + 3cP_o) + \frac{3}{2}cP_2 (2 + 3\tau^2)} \quad (27)$$

$$\frac{k_2^{*2}}{k_2^2} = \frac{(1 + 9cP_1)\left\{1 + \frac{3}{2}cP_2 (2 + 3\tau_2)\right\}}{1 + \frac{3}{4}cP_2 (4 - 9\tau^2)} \quad (28)$$

$$k_1^* = \omega/c_1^*, \; k_2^* = \omega/c_2^*$$

c_1^* and c_2^* are the effective wave speeds of plane longitudinal and shear waves, respectively. c is the volume concentration of inclusions in the matrix.

The attenuation caused by scattering (to this order of approximation) can also be calculated using equations (17) and (26). First, we note that the scattering cross sections for incident P and S waves are

$$\Sigma_p = \frac{9v_o^2\varepsilon^4}{4\pi a^4} \left[P_o^2 + 3(1 + 2\tau^3) P_1^2 + 5(1 + \frac{3}{2}\tau^5)P_2^2\right] \quad (29)$$

and

$$\Sigma_s = \frac{9v_o^2\varepsilon^4\tau}{4\pi a^4} \left[3P_1^2 (1 + 2\tau^3) + \frac{15}{4}\tau^2P_2^2 (1 + \frac{3}{2}\tau^5)\right] \quad (30)$$

where v_o is the volume of a spherical inclusion. Using (29) and (30), the attenuation coefficients α_p and α_s are then given by

$$\frac{\alpha_p}{k_1} = 3c\varepsilon^3 \left[P_o^2 + 3P_1^2(1 + 2\tau^3) + 5P_2^2(1 + 3\tau^5/2)\right] = Q_p^{-1} \quad (31)$$

$$\frac{\alpha_s}{k_2} = 3c\varepsilon^3 \left[3P_1^2 (1 + 2\tau^3) + \frac{15}{4}\tau^2P_2^2(1 + 3\tau^5/2)\right] = Q_s^{-1} \quad (32)$$

We see from equations (27) - (32) that the effective phase velocities and attenuation coefficients can be modified substantially by the thin layer because they depend on P_0 and P_2, which may be influenced strongly by this layer. Note that P_1 is independent of this layer because in our approximation we neglect its inertia.

As an example, consider a composite made up of SiC spherical particles in aluminum. The relevant physical properties of SiC are

$$\lambda_1 + 2\mu_1 = 474.2 \text{ GPa}, \quad \mu_1 = 188.1 \text{ GPa}$$

$$\rho_1 = 3.181 \text{ g/cm}^3, \quad \sigma_1 = 0.17.$$

Those of aluminum are

$$\lambda_2 + 2\mu_2 = 110.5 \text{ GPa}, \quad \mu_2 = 26.5 \text{ GPa}$$

$$\rho_2 = 2.706 \text{ g/cm}^3, \quad \sigma_2 = 0.34.$$

Then we find that

$$P_0 = -\frac{0.19 - 0.26(h/a)}{1 + 0.37(h/a)}$$

$$P_2 = -\frac{0.04 + 0.02(h/a) - 0.02(h/a)^2}{1 + 1.55(h/a) + 0.59(h/a)^2}$$

P_1 in this case is 0.02. Thus, for small h/a it is seen that the percentage changes in P_0 and P_2 are approximately 174(h/a)% and 88(h/a)%, respectively. These changes should be discernible in phase-velocity measurements.

REFERENCES

1. P.C. Waterman and R. Truell, J. Math. Phys., 2(1961) 512-537.
2. N. Yamakawa, Geophys. Mag., 31(1962) 97-103.
3. J.G. Fikioris and P.C. Waterman, J. Math. Phys., 5(1964) 1413-1420.
4. A.K. Mal and L. Knopoff, J. Inst. Math. Appl., 3(1967) 376-387.
5. A.K. Mal and S.K. Bose, Proc. Camb. Phil. Soc., 76(1974) 587-600.
6. G.T. Kuster and M.H. Toksöz, Geophysics, 39(1974) 587-606, 607-618.
7. S.K. Datta, J. Appl. Mech., 44(1977) 657-662.

8. S.K. Datta, Continuum Model of Discrete Systems (J.W. Provan, ed.), Univ. of Waterloo Press, 1978, pp. 111-127.
9. S.K. Datta, Continuum Models of Discrete Systems (E. Kröner and K. Anthony, eds.), Univ. of Waterloo Press, 1980, pp. 565-582.
10. C.M. Sayers, J. Phys. D: Appl. Phys., 13(1980) 179-184.
11. J.G. Berryman, J. Acoust. Soc. Am., 68(1980) 1809-1819.
12. J.G. Berryman, J. Acoust. Soc. Am., 68(1980) 1820-1831.
13. C.M. Sayers, J. Phys. D: Appl. Phys., 14(1981) 413-420.
14. C.M. Sayers and R.L. Smith, J. Phys. D: Appl. Phys., 16(1983) 1189-1194.
15. J.E. Gubernatis and E. Domany, Effects of Microstructure on the Speed and Attenuation of Elastic Waves, Report LA-UR-82-2630, Los Alamos National Laboratory, Los Alamos, 1982, 16 pp.
16. J.E. Gubernatis and E. Domany, Effects of Microstructure on the Speed and Attenuation of Elastic Waves, Report LA-UR-83-2611, Los Alamos National Laboratory, Los Alamos, 1983, 34 pp.
17. S.K. Datta and H.M. Ledbetter, Wave Propagation in Inhomogeneous Media and Ultrasonic Nondestructive Evaluation (G.C. Johnson, ed.), The American Society of Mechanical Engineers, New York, 1984, pp. 123-139.
18. H.M. Ledbetter and S.K. Datta, Materials Sci. Eng., 67(1984) 25-30.
19. H.M. Ledbetter, S.K. Datta, and R.D. Kriz, Acta Metall., 32(1984) 2225-2231.
20. V.K. Varadan, Y. Ma, and V.V. Varadan, J. Acoust. Soc. Am. 77(1985), 375-385.
21. C.M. Sayers, Wave Motion, 7(1985) 95-104.

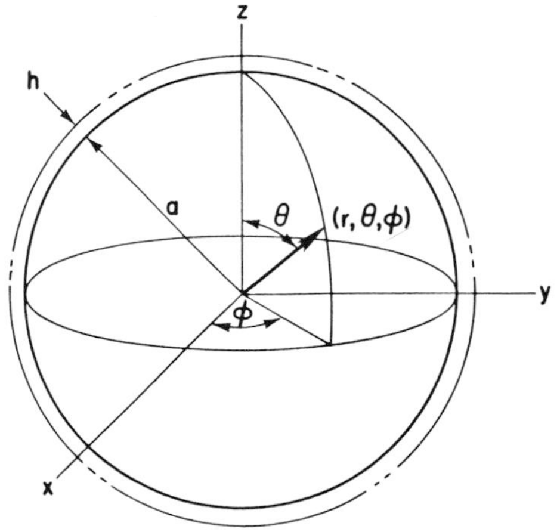

Fig. 1 Schematic diagram showing spherical polar coordinates for a spherical particle of radius a surrounded by a layer of thickness h.

STABILITY OF A LONG CYLINDRICAL SHELL WITH PARTIAL CONTACT AT THE END FACE

A. AZARKHIN[1] and J. R. BARBER[2]

[1]Product Eng. Div., Aluminum Company of America, ALCOA Tech. Center, PA 15069, USA, formerly at the University of Michigan.

[2]Dept. of Mech. Eng. and Applied Mech., University of Michigan, Ann Arbor, MI 48104, USA.

ABSTRACT

We study the partial contact of a long circular cylindrical shell pressed eccentrically by its end face against a rigid half-space, using the theory of thin shells in bending. The elastic displacements are presented in the form of a Fourier expansion, the coefficients of which are found by the variational formulation combined with the penalty method to enforce contact conditions. Results are applied to buckling of a centrally and eccentrically compressed bar, and a bar with an initial inperfection. These solutions are compared with the corresponding results of the strength of materials.

INTRODUCTION

Consider a circular semi-infinite cylinder of radius R and thickness t, one end of which is at rest on a half-space. The cylinder transmits an axial force P applied with some eccentricity e to the half-space. This description applies to a set of related problems. Suppose for example, that the cylinder is compressed centrally at the start. If the force P is less than the Euler critical load for a built-in bar

$$P_{cr} = \frac{\pi^2 EI}{4L^2} \qquad (1)$$

there is only the trivial zero solution for lateral displacements. Otherwise lateral displacements develop, causing some eccentricity of the loading at the lower end face (see figure 1). The

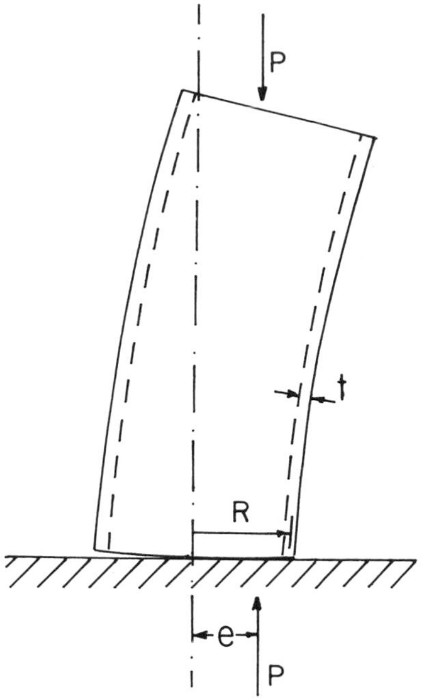

Figure 1. Geometry of the problem.

elementary strength of materials solution gives the relation for the stable equilibrium of a built-in bar (ref. 1):

$$kL = K(p) = \int_0^{\pi/2} \frac{d\phi}{\sqrt{1-p^2\sin^2\phi}} \qquad (2)$$

where $p = \sin(\alpha/2)$, $k = \sqrt{P/EI}$, α is the rotation of the upper end, L is the length of the bar and K(p) is the complete elliptic integral of the first kind. This solution is valid if the dimensionless eccentricity of the load

$$\frac{e}{R} = \left(\frac{2p}{KL}\right)\left(\frac{L}{R}\right) \leq 0.5. \qquad (3)$$

Otherwise the solution predicts tensile tractions, indicating that separation will occur.

To proceed with the stability problems of this type for $e/R \geq 0.5$, we first establish non-linear relationships between the eccentricity of the load and the average rotation of the end face. The corresponding curves for bars with different ratios R/t are obtained by the use of the theory of thin shells in bending. We then consider some examples, including that stated above, and show how the elementary solutions should be modified to take account of the partial contact at the end face.

CONTACT OF A CYLINDRICAL SHELL AND A HALF-SPACE

The problem of a thin-walled cylindrical shell in partial contact with a rigid plane has already been solved in ref. 2 and the solution and results are therefore only briefly summarized here. The hollow circular cylinder is treated as a thin-walled shell in bending. As in ref. 3, the elastic displacements are sought as a Fourier series, each term of which has the form $\sin(mx)f_m(y)$ or $\cos(mx)g_m(y)$ where x and y are respectively circumferential and longitudinal coordinates, and $f_m(y)$, $g_m(y)$ are combinations of trigonometric and exponential functions with four unknown coefficients for each m. The governing equations are uncoupled among m and satisfied by the choice of the form. The unknown coefficients have to be found from the contact boundary conditions. The following three boundary conditions are global and uncoupled among m:

(a) absence of the axial bending moment,

(b) absence of the effective shearing tractions,

(c) absence of the effective tangential tranctions.

These allow us to eliminate three out of four unknown coefficients for each m. The remaining coefficients are then found from the non-linear contact boundary conditions following from

(d) absence of tensile pressure in the region of contact,

(e) absence of relative displacements of the two bodies in the region of contact,

(f) non-penetration of the two bodies out of contact,

(g) absence of normal tractions out of contact.

For a numerical implementation, the problem was stated in the form of minimization for the functional of the potential energy of elastic deformation. The inequality conditions (e) and (f), which serve to determine the contact area, were enforced by the exterior penalty approach which has a clear mechanical interpretation. We introduced some artificial springs with a large stiffness on the surface, which were effective if there was overlapping of the two bodies, and otherwise ineffective. Since the elastic energy of the springs was included in the functional, the interpenetration of the bodies was discouraged in the process of minimization. The functional was minimized by giving small increments to all unknown coefficients in turn until the minimum with some prescribed allowance was reached.

Results are given in ref. 2 for a cylindrical shell with various geometrical parameters R/t and Poission's ratio $\nu = 0.3$. The results are condensed in figures 2 and 3. In particular, the relationship between the load eccentricity, e/R, and the extent of the area of contact is presented in figure 2.

An approximate solution to the problem can be obtained by "unwrapping" the cylinder, to give the plane stress problem of periodic patch-like contact between an elastic body with a wavy surface and a rigid plane. This approximation is used for a related thermoelastic problem by Burton et al (ref. 4). It is of interest to know how the shell thickness ratio, R/t, affects the accuracy of this approximation - i.e. the extent to which shell

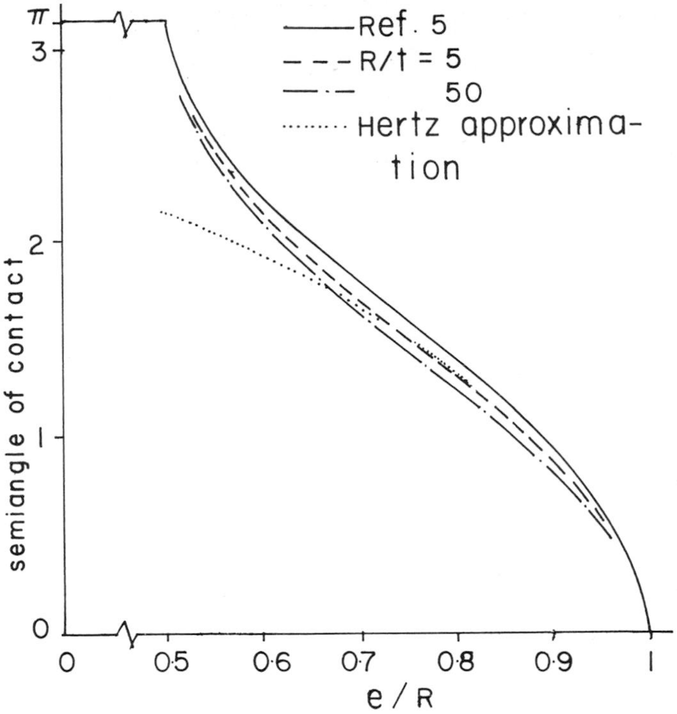

Figure 2. Influence of the thickness parameter R/t on the relationship between the eccentricity of the load, e/R, and the extent of the area of contact.

bending influences the contact problem.

An elegant complex variable solution for the contact of a sinusoidally wavy surface and a plane was given by Westergaard (ref. 5). The stress function, the contact distribution and the extent of the periodic contact patches were all obtained in closed form. Results from ref. 5 are shown in figure 2, from which we see that the unwrapping approximation slightly overestimates the area of contact. The error increases with increasing R/t, as we should expect, since the solution tends to that of a membrane shell for which the St. Venant principle fails (see discussion of this issue in the article 1 of the next section). The Hertzian approximation to the plane stress solution is also shown in Figure 2 for comparison.

Shell bending has a more noticeable effect on the flexibility of the shell, as can be seen from figure 3, which shows the angle shell end surface as a function of load eccentricity.

STABILITY PROBLEMS WITH PARTIAL CONTACT

The curves in figures 2 and 3 were obtained using the small deformation elastic shell theory, but it is nonetheless possible to use them for problems involving large displacements if the cylinder is long. This is justified because the partial contact solution merely provides a self-equilibrated correction to the strength of materials solution in the vicinity of the compressed end. Its effects will therefore only be significant near this end and the displacements there will be small, provided that the angle of relative rotation of the half-space and the mean end face of the cylinder is sufficiently small.

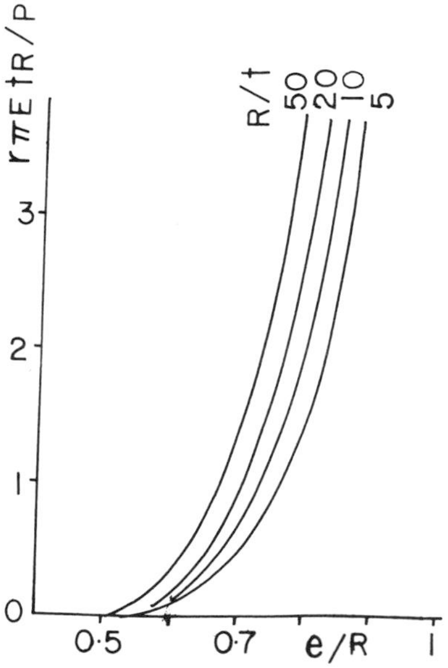

Figure 3. Influence of the thickness parameter R/t on the relationship between the eccentricity of the load e/R, and the rotation of the end face.

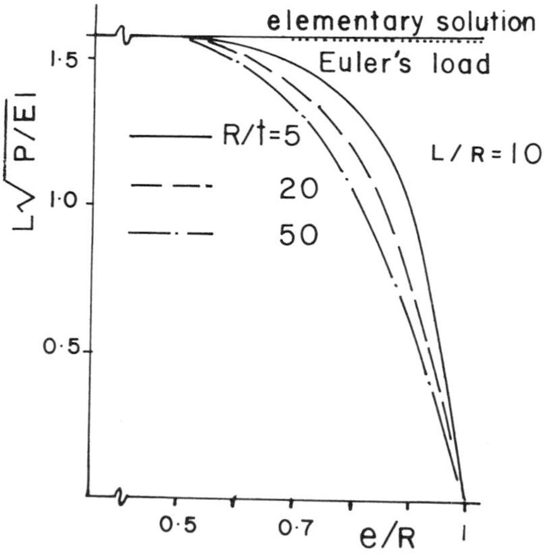

Figure 4. Influence of the thickness parameter R/t on the critical value of the force.

1. Centrally compressed bar.

We now return to the example of figure 1 and show how the solution should be modified for the case of the partial contact at the end face. The difference is that there is some angle of rotation of the lower end, due to local compression of the shell, and therefore the lower limit of integration in (1) is not zero. Equation (2) should be replaced with

$$kL = \int_{\phi_1}^{\pi/2} \frac{d\phi}{\sqrt{1-p^2 \sin^2\phi}} \tag{4}$$

where ϕ_1 is related to the rotation r of the lower end by (ref. 1)

$$\phi_1 = \sin^{-1}\left[\frac{1}{p} \sin\left(\frac{r}{2}\right)\right] \tag{5}$$

The expression (3) for the dimensionless eccentricity we replace with

$$\frac{e}{R} = \frac{w}{R} = \frac{2p}{kL}\left(\frac{L}{R}\right) \cos\phi_1 \tag{6}$$

where w is the lateral displacement of the upper end of the bar.

The angle of rotation r of the lower end and therefore ϕ_1, are unknown but they can be found by iteration. Suppose we assign some value to $r < \alpha$, and find the corresponding value of ϕ_1 from (5), and then kL from (4) and e/R from (6). The latter gives the end rotation r from figure 3. If this differs from the value of r we started with, we try another initial value.

This algorithm was used to obtain numerical results which are shown in figures 4, 5. Figure 4 shows the relation between the load P and the eccentricity e for L/R = 10 and various values of R/t. One remark about figure 4 should be made. We can write

$$P = \frac{EI}{L^2} f(\frac{e}{R}, \frac{R}{t}) = \frac{\pi E t R^3}{L^2} f(\frac{e}{R}, \frac{R}{t}) \quad (7)$$

where the function f is given by the curves of figure 4. The load P depends on t explicitly in (7), but we can state more than that. We note from fig. 4 that the function f decreases with increasing R/t, therefore the load drops faster than the thickness t. The explanation is as follows. For a sufficiently large ratio R/t the stress state of the bar tends to that of a

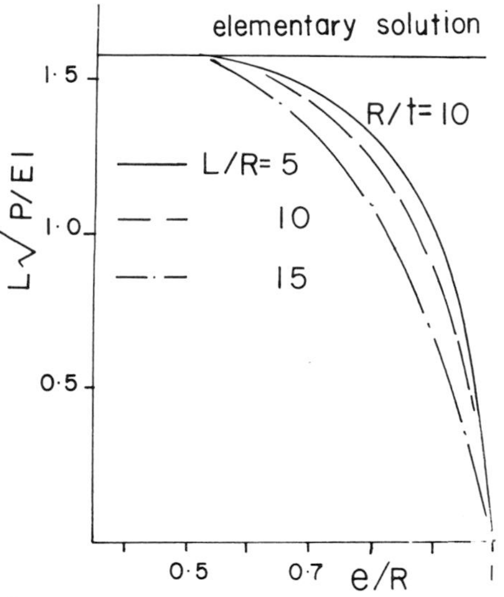

Figure 5. Influence of the length parameter L/R on the critical value of the force.

membrane shell. Equilibrium equations for the membrane shell are statically determinate (ref. 3), and it follows that the axial stress N_y does not depend on y, i.e. the stress state does not tend to that of the strength of materials. St. Venant principle fails and the bar loses distributing ability which decreases its resistance.

Figure 5 relates the load to the eccentricity when R/t is fixed and L/R varies. The elementary solution assuming complete contact is also shown for comparison in both figures 4 and 5. As we should expect, the difference is substantial. With complete contact the load grows slightly as the lateral displacement of the upper end (and therefore eccentricity of the load at the lower end) increases. With partial contact, the load drops as the eccentricity grows. The force P becomes equal to zero when $e/R = 1$. This point in the fig. 4 and 5 corresponds to the limit configuration when the bar topples over.

We can see that the difference between the elementary solution (2) which takes into account large displacements and the Euler formula $kL = \pi/2$ is hardly noticeable in the region of interest. For this reason we shall use the assumption of small displacements in the following problems.

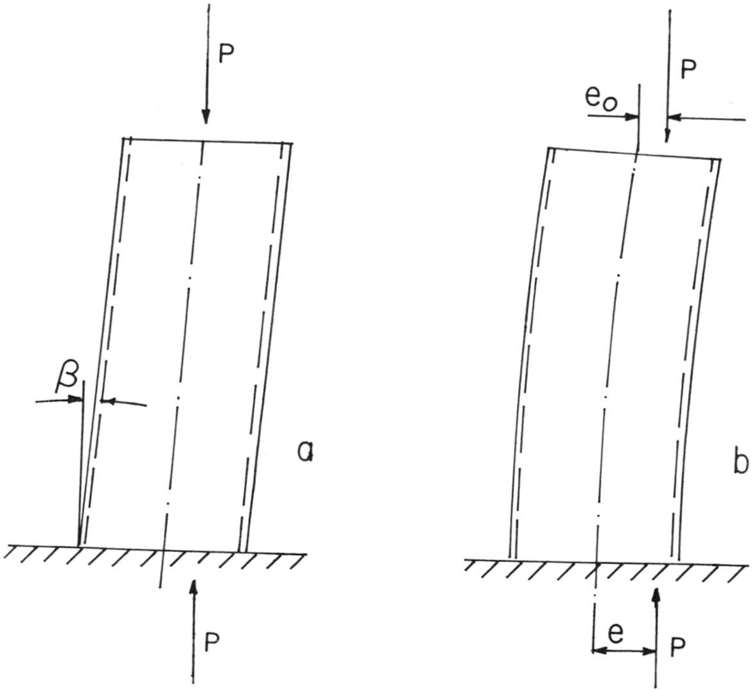

Figure 6. Inaccuracy of manufacture. (a) Initial misalignment of the bar. (b) Initial displacement of the force.

2. Effect of initial imperfection.

We note that the model with partial contact does not change

the critical load given by (1) and figures 4 and 5 apply if the axial displacement of the upper end is controlled. However, the structure is sensitive to inaccuracy of manufacture and a relatively small imperfection can reduce the critical load considerable. Suppose, for example, that the lower end of the cylinder is not quite perpendicular to the axis and the position of the cylinder deviates slightly from the vertical line by a small angle β (see fig. 6a). In contrast to the previous problem, the the force P is small. Generally, the eccentricity depends on the deformation of the bar and the rotation of the lower end face. Both are unknown in advance but can be found by iteration. The algorithm remains almost the same, but we should keep in mind that the angle of relative rotation of the end face and the half-space which is found from fig. 3, and the angle r introduced in (5) differ by the amount β.

Results for several values of β are given in figure 7. The load P, which is applied at the center of the upper end of the cylinder, already has an eccentricity $e = \beta L$ relative to the lower end. This increases non-linearly as P increases up to a maximum which corresponds to the onset of instability. The remaining portions of the curves correspond to unstable equili-

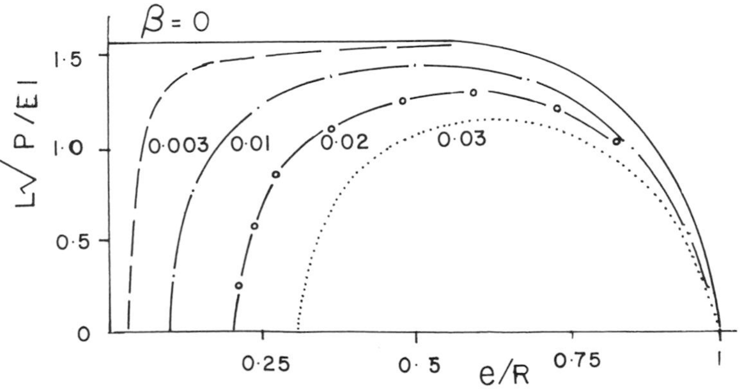

Figure 7. Influence of initial misalignment of the bar.

brium and could only be realized if the displacement of the upper end face were controlled. In each case partial contact occurs when e exceeds R/2. Figure 7 demonstrates the influence of the accuracy of the structure on the critical load. If for example B = 0.02 (or 1.15^{o}), the limit load is reduced by 17%. The influence would be even more pronounced for L/R>10.

3. Effect of a non-central load.

Finally, we consider the case where the force P is displaced from the center as shown in figure 6b. The governing equation

$$EIw^{IV} - Pw'' \tag{8}$$

has the solution

$$\frac{w}{R} = \left[\frac{e_o}{R\cos(kL)} + \frac{r}{(kL)}\left(\frac{L}{R}\right)\tan(kL)\right][1-\cos(ky)] + \frac{r}{(kL)}\left(\frac{L}{R}\right)\sin(ky) \tag{9}$$

Equation (9) satisfies all boundary conditions of the problem by choice of the form, namely the shearing force is equal to zero at y = 0 and y = L, w = 0 at y = 0, and the bending moment $M = Pe_o$ at y = L. Putting y = L in (9), we obtain

$$\frac{e}{R} = \frac{e_o}{R\cos(kL)} = \frac{r\cdot\pi EtR}{P}(kL)\left(\frac{R}{L}\right)\tan(kL) \tag{10}$$

For any given kL, equation (10) has two unknowns: e and r. Another relation between these quantities is defined by figure 3. This "system of equations" was solved by a procedure similar to that described before. Numerical results for e_o/R = 0, 0.05 and 0.1 with R/t = 10 and L/R = 10 are given in figure 8. We can see, for example, that if the force is displaced by a distance equal to 5% of the radius, the critical load is reduced by 11%.

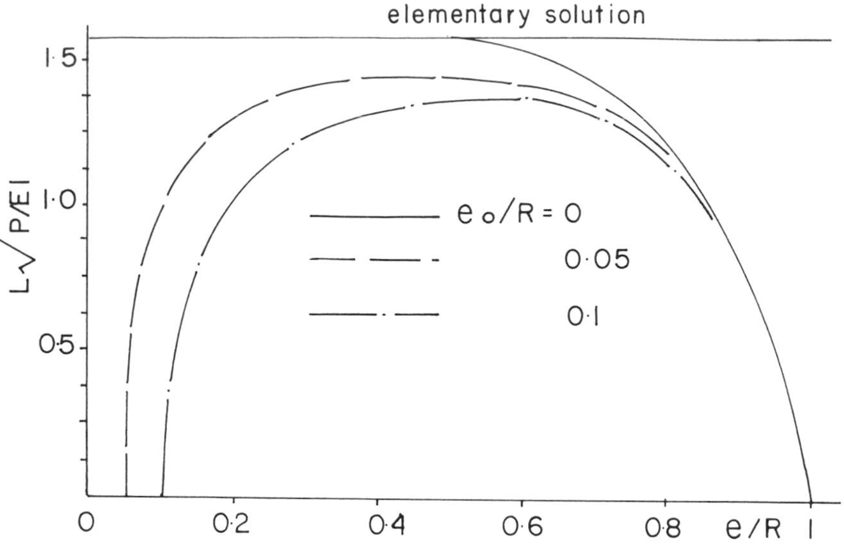

Figure 8. Influence of the initial displacement of the force.

It is not surprising that figures 7 and 8 are qualitatively similar. Although the two problems are not identical, they approach the same solution as imperfectness tends to zero.

A number of other problems might be treated in the same way. For example, one can treat the stability of a bar subjected to its own weight, or with initial curvature, or any combination of load and imperfections mentioned in this section etc.

REFERENCES

1. S. Timoshenko, Theory of elastic stability, 2nd edn., McGraw-Hill, N.Y., 1961, 541 pp.

2. A. Azarkhin and J. R. Barber, On partial contact of a thin-walled circular cylinder and a rigid half-space, Int. J. Mech. Sci. (in press).

3. W. Flugge, Stresses in shells, 2nd edn., Springer-Verlag, 1973, 525 pp.

4. R. A. Burton, V. Nerlikar and S. R. Kilaparti, Thermoelastic instability in a seal-like configuration, Wear, 24 (1973), 177-188.

5. H. M. Westergaard, Bearing pressures and cracks, J. Appl. Mech., 6(1939), 49-53.

INTERFACE PROBLEMS IN GEOMECHANICS

UNILATERAL CONTACT AT AN INTERNALLY INDENTED SMOOTH ELASTIC INTERFACE

A.P.S. SELVADURAI
Department of Civil Engineering, Carleton University, Ottawa, Ontario, Canada K1S 5B6

ABSTRACT

The present paper examines the problem of the interface separation at a transversely isotropic elastic, smoothly precompressed interface. The separation is induced by a disc shaped rigid inclusion of finite thickness. The problem serves as an approximate mathematical model of the situation where separation of a pre-factured interface, in a resource bearing stratum, is induced by the "proppant material" of a drilling fluid. The analysis of the problem focusses on the evaluation of the radius of the zone of separation beyond the proppant region. The numerical results presented in the paper illustrate the manner in which the radius of the zone of separation is influenced by the magnitude of the precompression and the elastic deformability characteristics of the medium.

INTRODUCTION

The study of unilateral contact between elastic solids is an important aspect of the mathematical theory of elasticity. Comprehensive accounts of current developments in this area are given by Dundurs and Stippes (ref. 1), dePater and Kalker (ref. 2), Duvaut and Lions (ref. 3) and Gladwell (ref. 4). Unilateral contact problems which deal with the interaction of either an elastic halfspace region and elastic layers or elastic layers and rigid boundaries have been examined by Keer and Chantaramungkorn (ref. 5), Keer et al. (ref. 6) and Tsai et al. (ref. 7). These authors have examined the unilateral contact between the various configurations of elastic halfspaces and elastic layers, which is induced by localized concentrated forces and rigid indentors. Similarly, Pu and Hussain (ref. 8), Civelek and Erdogan (ref. 9) and Gecit and Erdogan (ref. 10) have examined problems which relate to separation at the interface between an elastic layer and a rigid boundary. Similar problems pertaining to elastic layers which are subjected to gravitational forces are considered by Keer and Silva (ref. 11), Civelek and Erdogan (ref. 12) and Gecit (ref. 13). The papers by Comninou et al. (ref. 14), Schmueser et al. (refs. 15,16), and Selvadurai and Au (ref. 17) examine the class of unilateral contact problem in which frictional effects are present at the interface. Selvadurai (ref. 18) has examined the problem of the body force induced separation at a precompressed transversely isotropic elastic interface. Gladwell and Hara (ref. 19) have examined a more general problem which relates to the separation at a smoothly compressed bi-material elastic interface which is induced by a smoothly embedded axisymmetric inclusion with an oblate spheroidal shape. Selvadurai (ref. 20) has examined the problem of the interface separation by a rigid disc inclusion where the associated integral equations are solved in an approximate manner. Further studies involving tensionless contact at smooth elastic interfaces are given by Dundurs et al. (ref. 21). Finally, the category of problems that deals with unilateral contact between beams and plates resting in smooth contact with elastic media has been examined by Weitsman (ref. 22), Gladwell and Iyer (ref. 23), Gladwell (ref. 24) and Svec (ref. 25). References to further work in this area are also given by Selvadurai (ref. 26), Gladwell (ref. 4) and Panagiotopoulos (ref. 27).

This paper presents a solution to the problem of the interface separation at a transversely isotropic elastic precompressed interface which is induced by a disc shaped rigid circular inclusion. This particular problem is of interest to the examination of interface separation at a prefractured resource bearing formation which is induced by injected proppant materials (Figure 1). An assessment of the proppant induced zone of separation is important to the determination of the efficiency of the hydraulic fracturing endeavour (see e.g. Cleary ref. 28). The idealization of the proppant region by a rigid disc shaped inclusion gives an upper bound estimate for the zone of separation. The idealized problem of the disc inclusion induced separation at the precompressed elastic interface is illustrated in Figure 2. The precompression σ_v is assumed to be such that complete contact is maintained between the inclusion and the transversely isotropic elastic halfspaces. A zone of separation of radius b is created beyond the indenting inclusion. Smooth contact is maintained beyond the zone of separation. The extent of the separation zone is an unknown parameter in the problem which is influenced by the geometric aspect ratio of the inclusion (i.e. the thickness to diameter ratio 2h/2a or alternatively the volume of proppant), the ratio σ_v/c_{44} where c_{44} is the shear modulus of the transversely isotropic material and c_{ij}/c_{44} where c_{ij} are the elastic constants of the transversely isotropic elastic material. In order to determine the radius of separation, we utilize solutions to two auxiliary problems. These are three part mixed boundary value problems in which displacements are prescribed in the regions $r \in (0,a)$ and $r \in (b,\infty)$ and normal tractions are prescribed in the region $r \in (a,b)$. In each case, the analysis focusses on the evaluation of the stress intensity factor at the boundary $r = b$. The condition of a vanishing stress intensity factor at the point of separation is used to develop the characteristic equation required for the estimation of b. Numerical results presented in the paper illustrate the manner in which the radius of the zone of separation is influenced by σ_v/c_{44}, h/a and c_{ij}/c_{44}.

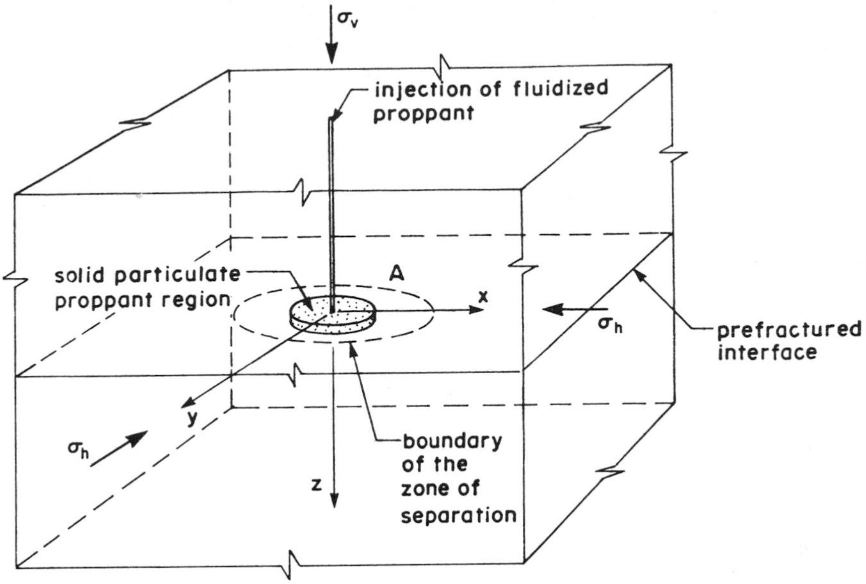

Figure 1: Proppant - induced separation in a resource bearing stratum

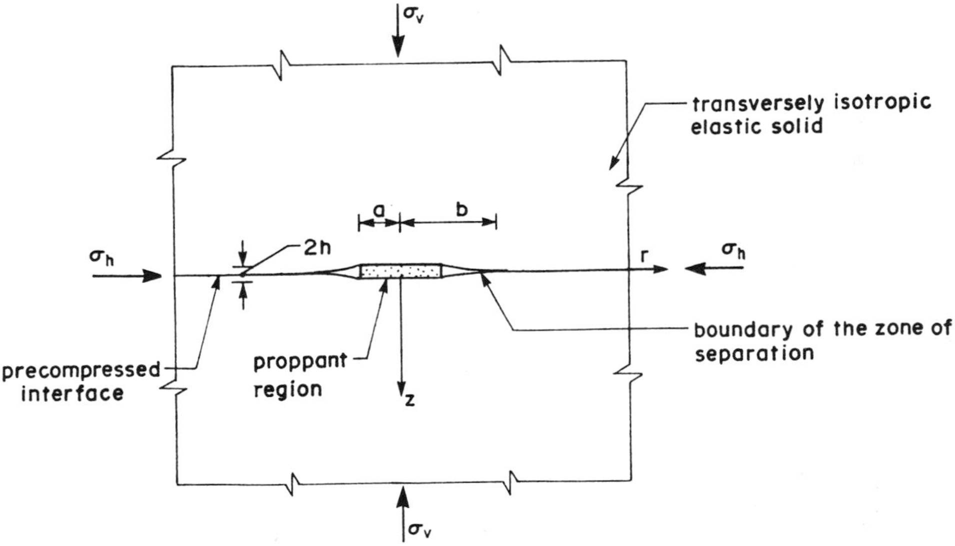

Figure 2: Detail at A

FUNDAMENTAL EQUATIONS

The methods of analysis of three-dimensional problems in transversely isotropic elastic materials make extensive use of the potential function techniques proposed by Elliott (refs. 29,30) and Lekhnitskii (ref. 31). Accounts of these developments are given by Green and Zerna (ref. 32) and Eubanks and Sternberg (ref. 33). It can be shown that in the absence of body forces, the axisymmetric displacement and stress fields can be expressed in terms of two functions $\phi_\alpha(r,z)$ ($\alpha = 1,2$) which are solutions of

$$\left(\frac{\partial^2}{\partial r^2} + \frac{1}{r}\frac{\partial}{\partial r} + \frac{\partial^2}{\partial z_\alpha^2}\right)\phi_\alpha(r,z) = 0 \tag{1}$$

where $z_\alpha = z/\sqrt{\nu_\alpha}$, and ν_1 and ν_2 are the roots of the equation

$$c_{11}c_{44}\nu^2 + \{c_{13}(2c_{44}+c_{13}) - c_{11}c_{33}\}\nu + c_{33}c_{44} = 0 . \tag{2}$$

The cylindrical polar coordinate system (r,θ,z) is chosen such that the z-axis is parallel to the axis of material symmetry. The roots of (2) may be real or complex depending upon the magnitude of the elastic constants c_{ij}. The displacement and stress fields in the transversely isotropic elastic material can be represented in terms of $\phi_\alpha(r,z)$. The particular displacement and stress components relevant to the separation problem can be written in the form

$$u_z(r,z) = \frac{\partial}{\partial z}\{k_1\phi_1 + k_2\phi_2\} \tag{3}$$

$$\sigma_{zz}(r,z) = (k_1c_{33} - \nu_1c_{13})\frac{\partial^2\phi_1}{\partial z^2} + (k_2c_{33} - \nu_2c_{13})\frac{\partial^2\phi_2}{\partial z^2} \tag{4}$$

$$\sigma_{rz}(r,z) = c_{44}\{(1+k_1)\frac{\partial^2\phi_1}{\partial r\partial z} + (1+k_2)\frac{\partial^2\phi_2}{\partial r\partial z}\} \tag{5}$$

where k_1 and k_2 are given by

$$k_\alpha = \frac{c_{11}\nu_\alpha - c_{44}}{c_{13} + c_{44}} \quad ; \quad (\alpha = 1,2) \tag{6}$$

THE UNILATERAL CONTACT PROBLEM

Consider the axisymmetric problem in which two transversely isotropic elastic halfspace regions are precompressed by a homogeneous stress field σ_v. The smooth contact between the halfspace regions is perturbed by a rigid circular disc inclusion of diameter 2a and thickness 2h. The wedging action creates a zone of separation of radius b at the smooth interface. The magnitude of σ_v is such that complete contact is maintained between the disc inclusion and the halfspace regions beyound $r \geq b$ and complete contact is maintained between the inclusion and the halfspace regions in the region $r \leq a$. The extent of the separation zone is obtained by evaluating the stress intensity factor K_I^* at the location $r = b^+$ (where the positive superscript denotes a point immediately beyound $r = b$) for the following auxiliary problems. The first problem considers the internal indentation of a penny-shaped crack (of radius b) located in a transversely isotropic elastic solid by a smooth rigid disc inclusion of thickness 2h (Figure 3). The relevant stress intensity factor is denoted by K_I^h. The second problem considers the evaluation of the stress intensity factor at the outer boundary $r = b$ of an annular crack which is subjected to a uniform internal tensile stress σ_v in the region $r \in (a,b)$ (Figure 4). The associated stress intensity factor is denoted by $K_I^{\sigma_v}$. The characteristic equation required for the determination of the radius of the zone of separation is given by the consistency condition $K_I^* = K_I^h + K_I^{\sigma_v} = 0$. The two auxiliary problems are illustrated in Figures 3 and 4, and the relevant mixed boundary conditions can be prescribed by focussing attention to a single halfspace region ($z \geq 0$).

For the symmetric interval indentation of the penny-shaped crack by a smooth disc inclusion we have

$$\sigma_{rz}(r,0) = 0 \; ; \quad r \geq 0 \tag{7}$$

$$u_z(r,0) = h \; ; \quad 0 \leq r \leq a \tag{8}$$

$$u_z(r,0) = 0 \; ; \quad b \leq r < \infty \tag{9}$$

$$\sigma_{zz}(r,0) = 0 \; ; \quad a < r < b \, . \tag{10}$$

For the tensile loading of the annular crack we have

$$\sigma_{rz}(r,0) = 0 \; ; \quad r \geq 0 \tag{11}$$

$$u_z(r,0) = 0 \; ; \quad 0 \leq r \leq a \tag{12}$$

$$u_z(r,0) = 0 \; ; \quad b \leq r < \infty \tag{13}$$

$$\sigma_{zz}(r,0) = \sigma_v ; \quad b < r < a \, . \tag{14}$$

For the purposes of examining these auxiliary problems, it is convenient to employ a Hankel transform development of the governing differential equation (1). The integral representation for $\phi_\alpha(r,z)$ can be chosen such that the displacement and stress fields satisfy the regularity conditions $u_i \to 0$ as $(r^2+z^2)^{\frac{1}{2}} \to \infty$ and $\sigma_{ij} \to 0$ as $(r^2+z^2)^{\frac{1}{2}} \to \infty$. The relevant solution is given by (ref. 30)

$$\phi_\alpha(r,z) = \int_0^\infty \xi A_\alpha(\xi) \exp\{- \frac{\xi z}{\sqrt{\nu_\alpha}}\} J_0(\xi r) d\xi \tag{15}$$

where $A_\alpha(\xi)$ are arbitrary functions which need to be determined by satisfying

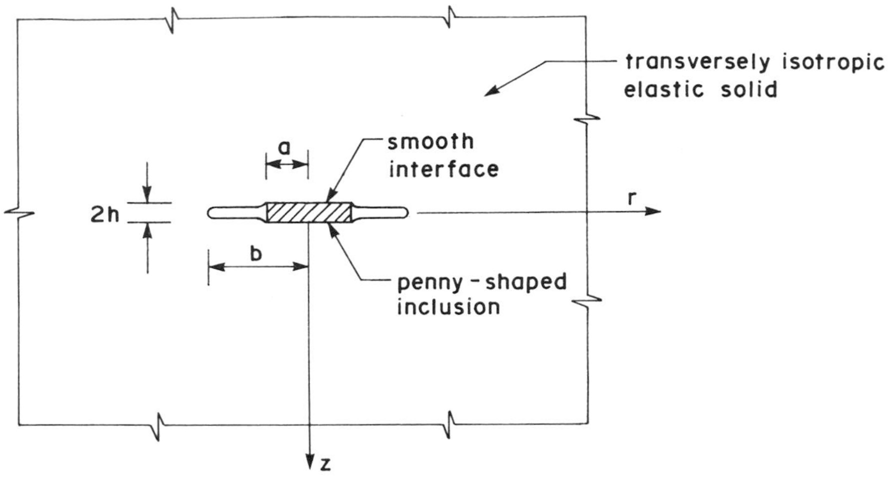

Figure 3: Internal indentation of the penny-shaped crack

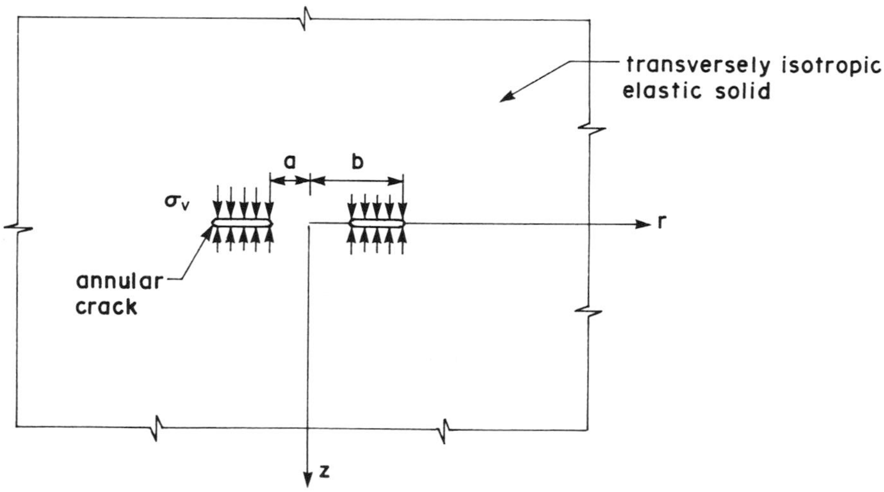

Figure 4: Uniform tensile loading of the annular crack

the mixed boundary conditions applicable to each auxiliary problem.

THE INTERNAL INDENTATION PROBLEM

The first auxiliary problem concerns the smooth internal indentation of the penny-shaped crack by a disc inclusion of finite thickness (Figure 3). Considering the solution of (1) given by (15) and the expressions for u_z, σ_{zz} and σ_{rz} given by (3) to (6), it can be shown that the mixed boundary conditions (7) to (10) yield the following system of triple integral equations for a single unknown function $A(\xi)$.

$$H_0[\xi A(\xi);r] = \frac{h(1+k_1)(1+k_2)}{(k_1-k_2)} \quad ; \quad 0 \le r \le a \tag{16}$$

$$H_0[\xi^2 A(\xi);r] = 0 \quad ; \quad a < r < b \tag{17}$$

$$H_0[\xi A(\xi);r] = 0 \quad ; \quad b \le r < \infty \tag{18}$$

where $H_n[\omega(\xi);r]$ is the Hankel operator of order n defined by

$$H_n[\omega(\xi);r] = \int_0^\infty \xi\omega(\xi)J_n(\xi r)d\xi. \tag{19}$$

The set of triple integral equations defined by (16) to (18) can be solved in a variety of ways and these procedures are summarized, among other, by Cooke (ref. 34), Williams (ref. 35), Tranter (ref. 36), Sneddon (ref. 37), Kanwal (ref. 38) and Selvadurai and Singh (ref. 39). Following refs. (34 and 39) we assume that (17) admits a representation of the form

$$H_0[\xi^2 A(\xi);r] = \begin{cases} f_1(r) \;; & 0 < r < a \quad (20) \\ f_2(r) \;; & b < r < \infty . \quad (21) \end{cases}$$

Using this representation, the equations (16) and (18) can be expressed in the following forms:

$$\int_0^a \eta f_1(\eta)L_1(\eta,r)d\eta + \int_b^\infty \eta f_2(\eta)L_1(\eta,r)d\eta = \frac{h(1+k_1)(1+k_2)}{(k_1-k_2)} \quad ; \quad 0 \le r \le a \tag{22}$$

$$\int_0^a \eta f_1(\eta)L_1(\eta,r)d\eta + \int_b^\infty \eta f_2(\eta)L_1(\eta,r)d\eta = 0 \quad ; \quad b \le r < \infty \tag{23}$$

where

$$L_1(\eta,r) = \int_0^\infty J_0(\xi\eta)J_0(\xi r)d\xi = H_0[\xi^{-1}J_0(\xi\eta);r] . \tag{24}$$

Also, the equations (22) and (23) yield the following system of integral equations for $f_1(\eta)$ and $f_2(\eta)$

$$f_1(\eta) = \frac{2}{\pi(a^2-\eta^2)^{\frac{1}{2}}}\left[\frac{\eta h(1+k_1)(1+k_2)}{(k_1-k_2)} - \int_b^\infty \frac{t(t^2-a^2)^{\frac{1}{2}} f_2(t)dt}{(t^2-\eta^2)}\right] \quad ; \; 0 \leq \eta \leq a \tag{25}$$

$$f_2(\eta) = \frac{2}{\pi(\eta^2-b^2)^{\frac{1}{2}}}\left[-\int_0^a \frac{t(b^2-t^2)^{\frac{1}{2}} f_1(t)dt}{(\eta^2-t^2)}\right] \quad ; \; b \leq \eta < \infty . \tag{26}$$

By substituting the value of $f_2(\eta)$ defined by (26) into (25) and making use of the substitutions

$$\psi(\tilde{\eta}) = \frac{\pi a(k_1-k_2)}{2h(1+k_1)(1+k_2)}(1-\tilde{\eta}^2)^{\frac{1}{2}} f_1(\tilde{\eta})$$

$$\eta = \tilde{\eta}a \quad ; \quad \xi = \tilde{\xi}a \tag{27}$$

we obtain the following Fredholm integral equation of the second-kind for $\psi(\tilde{\eta})$:

$$\psi(\tilde{\eta}) = 1 + \int_0^1 \psi(\tilde{\xi})K(\tilde{\xi},\tilde{\eta})d\tilde{\xi} \quad ; \quad 0 \leq \tilde{\xi} \leq 1 \tag{28}$$

where

$$K(\tilde{\xi},\tilde{\eta}) = -\frac{2\tilde{\xi}(1-c^2\tilde{\xi}^2)^{\frac{1}{2}}}{\pi^2(1-\tilde{\xi}^2)^{\frac{1}{2}}}\,[\Phi(\tilde{\xi}) - \Phi(\tilde{\eta})] \tag{29}$$

$$\Phi(\gamma) = \zeta\,\ell n\{\frac{1+c\zeta}{1-c\zeta}\} \quad ; \; \zeta = \left[\frac{1-\gamma^2}{1-c^2\gamma^2}\right]^{\frac{1}{2}} \quad ; \; (\gamma = \tilde{\xi} \text{ or } \tilde{\eta}) \tag{30}$$

and $c = a/b$. The expression for $\lim_{\tilde{\xi}\to\tilde{\eta}}[K(\tilde{\xi},\tilde{\eta})]$ can be found by invoking L'Hospital's rule. The resulting limit can be expressed as

$$\lim_{\tilde{\xi}\to\tilde{\eta}}[K(\tilde{\xi},\tilde{\eta})] = C(\tilde{\eta}) = \frac{(1-c^2)\tilde{\eta}\Omega(\tilde{\eta})}{\pi^2(1-c^2\tilde{\eta}^2)(1-\tilde{\eta}^2)} \tag{31}$$

with

$$\Omega(\tilde{\eta}) = \frac{2\rho}{(1-\rho^2)} + \ell n(\frac{1+\rho}{1-\rho}) \quad ; \; \rho = c\left[\frac{1-\tilde{\eta}^2}{1-c^2\tilde{\eta}^2}\right]^{\frac{1}{2}} . \tag{32}$$

The analysis of the internal indentation problem related to the penny-shaped crack is thus reduced to the solution of the Fredholm integral equation of the second-kind defined by (28). This equation can be solved in a variety of ways (see e.g. Atkinson (ref. 40) and Baker (ref. 41)). In ref. (39) the equation is solved by assuming that $\psi(\tilde{\eta})$ admits a power series expansion of the form

$$\psi(\tilde{\eta}) = \sum_{i=1}^{n} c^i \psi_i(\tilde{\eta}). \tag{33}$$

Alternatively, a Gaussian quadrature technique (ref. 42) can be used to solve the integral equation (28).

The result of primary interest to the study of the unilateral contact problem is the stress intensity factor K_I^h at the boundary of the penny-shaped crack. The stress intensity factor at the crack boundary is given by

$$K_I^h = \lim_{r \to b^+} [2(r-b)]^{\frac{1}{2}} \sigma_{zz}(r,0) . \tag{34}$$

Using the power series expansion scheme, equation (34) can be evaluated in the form

$$K_I^h = \frac{4ch}{\pi^2 \sqrt{b}} \quad \Phi N(c) \tag{35}$$

where

$$\Phi = \left\{ \frac{\sqrt{\nu_2}(k_1 c_{33} - \nu_1 c_{13})(1+k_2) - \sqrt{\nu_2}(k_2 c_{33} - \nu_2 c_{13})(1+k_1)}{\sqrt{\nu_1 \nu_2}\ (k_1 - k_2)} \right\} \tag{36}$$

$$N(c) = 1 + \frac{4c}{\pi^2} + c^2\{\frac{1}{3} + \frac{16}{\pi^4}\} + c^3\{\frac{20}{9\pi^2} + \frac{64}{\pi^6}\}$$

$$+ c^4\{\frac{112}{9\pi^4} + \frac{256}{\pi^8} + \frac{1}{5}\} + O(c^5) \cdot \tag{37}$$

Using the quadrature scheme, the result for K_I^h can be computed from the result

$$K_I^h = \frac{4c\Phi}{\pi^2\sqrt{b}} \int_0^1 \frac{\tilde{\xi}\psi(\tilde{\xi})d\tilde{\xi}}{(1-\tilde{\xi}^2)^{\frac{1}{2}}(1-c^2\tilde{\xi}^2)^{\frac{1}{2}}} \quad . \tag{38}$$

The results derived from both schemes show excellent agreement for $c \in (0,0.6)$ (ref. 42).

TENSILE LOADING OF THE ANNULAR CRACK

The second auxiliary problem concerns the uniform tensile loading of a plane annular crack (Figure 4) located in a transversely isotropic elastic solid. Using the solution (15) and the expressions for u_z, σ_{zz} and σ_{rz} given by (3) to (6), it can be shown that the mixed boundary conditions (11) to (14) yield the following system of triple integral equations for an unknown function $A^*(\xi)$:

$$H_o[\xi A^*(\xi);r] = 0 \qquad ; \quad 0 \le r \le a \tag{39}$$

$$H_o[\xi^2 A^*(\xi);r] = \frac{\sigma_v}{c^*} \qquad ; \quad a < r < b \tag{40}$$

$$H_o[\xi A^*(\xi);r] = 0 \qquad ; \quad b \le r < \infty \tag{41}$$

where c* is a constant which depends solely on the elastic constants of the transversely isotropic elastic solid. The system of triple integral equations defined by (39) to (41) can be analysed by adopting the various techniques outlined in the refs.(34-38). The annular crack problem related to the isotropic elastic solid has been examined by several authors (refs. 43-46). More recently, Selvadurai and Singh (ref. 47) have applied the power series expansion technique for the solution of the annular crack problem related to an isotropic elastic solid. These techniques are directly applicable to the transversely isotropic case. The details of the analytical procedures will not be pursued here. It is sufficient to record the expression for the stress intensity factor $K_I^{\sigma_v}$ at r = b; we have

$$K_I^{\sigma_v} = -\frac{2\sigma_v\sqrt{b}}{\pi} M(c) \tag{42}$$

where

$$\begin{aligned} M(c) = {} & 1 - \frac{4}{\pi^2}c - \frac{16}{\pi^4}c^2 - c^3\{\frac{1}{8} + \frac{64}{\pi^6}\} \\ & + c^4\{\frac{16}{3\pi^4} + \frac{4}{\pi^2}(\frac{1}{24} - \frac{8}{9\pi^2} + \frac{64}{\pi^6} + \frac{4}{9\pi^3})\} \\ & + c^5\{\frac{16}{\pi^4}(\frac{1}{24} + \frac{64}{\pi^6} - \frac{8}{9\pi^3} + \frac{8}{9\pi^2}) + \frac{256}{9\pi^2} - \frac{4}{15\pi^2}\} \\ & + O(c^6) . \end{aligned} \tag{43}$$

THE SEPARATION ZONE

The analysis of the unilateral contact problem focusses on the evaluation of the radius of the zone of separation. The condition

$$K_I^h + K_I^{\sigma_v} = 0 \tag{44}$$

gives the characteristic equation for the ratio b/a. For example, using the results for K_I^h and $K_I^{\sigma_v}$ which are evaluated in power series form, we can rewrite (44) in the form

$$S_1(\lambda) - (\frac{2\sigma_v}{c_{44}})(\frac{a}{h})S_2(\lambda) = 0 \tag{45}$$

where

$$\begin{aligned} S_1(\lambda) = \Phi \Big[& \frac{4\lambda^4}{\pi} + \frac{16}{\pi^3}\lambda^3 + \lambda^2(\frac{64}{\pi^5} + \frac{4}{3\pi}) + \lambda(\frac{80}{9\pi^3} + \frac{256}{\pi^7}) \\ & + (\frac{448}{9\pi^5} + \frac{1024}{\pi^9} + \frac{4}{5\pi}) \Big] \end{aligned} \tag{46}$$

$$S_2(\lambda) = \Big[\lambda^6 - \frac{4}{\pi^2}\lambda^5 - \frac{16}{\pi^4}\lambda^2 - \lambda^3\left(\frac{1}{8} + \frac{64}{\pi^6}\right)$$

$$- \lambda^2\left\{\frac{16}{3\pi^4} + \frac{4}{\pi^2}\left(\frac{1}{24} - \frac{8}{9\pi^2} + \frac{64}{\pi^6} + \frac{4}{9\pi^3}\right)\right\}$$

$$- \lambda\left\{\frac{16}{\pi^4}\left(\frac{1}{24} + \frac{64}{\pi^6} - \frac{8}{9\pi^3} + \frac{8}{9\pi^2}\right) + \frac{256}{9\pi^6} - \frac{4}{15\pi^2}\right\}\Big] \qquad (47)$$

and

$$\lambda = \frac{b}{a}\ . \qquad (48)$$

The radius of the zone of separation is given by the lowest root of (45) which satisfies the constraint $\lambda > 1$. The characteristic equation (45) can be solved numerically for various values of $(2\sigma_v/c_{44})$ and (a/h) to generate the relevant results.

NUMERICAL RESULTS AND CONCLUSIONS

The Figures 5 and 6 represent the manner in which the extent of the separation zone b/a is influenced by $(2\sigma_v/c_{44})$, (a/h), the material isotropy and transverse isotropy of the precompressed regions. The results for both classes of materials show consistent trends. The particular form of transverse isotropy corresponds to $\nu_{hh} = \nu_{vh} = 0.2$, $G_{vh}/E_v = 0.4$ and $E_h/E_v = 9$ where the suffixes v and h refer to the vertical and horizontal principal directions. As the parameter $2\sigma_v/c_{44}$ tends to unity (i.e., the overburden stress approaches one half the shear modulus of the material) the value of b/a approaches unity. Since these results have been derived by using the series representations for K_I^h and $K_I^{\sigma_v}$, the accuracy of the solution diminishes as λ (or c) approaches unity. The results can be considered accurate for all values of $b/a > 1.5$. It is evident that the axial precompression has a significant influence on the zone of separation. In terms of the application of the results to the separation at a prefractured geological interface it is evident that a substantial region of the interface will experience separation when the fracture is located at shallow depths. Owing to the penny-shaped nature of the indenting inclusion region the estimates for the separation region would represent an upper bound. The presence of transverse isotropy enhances the extent of the separation zone.

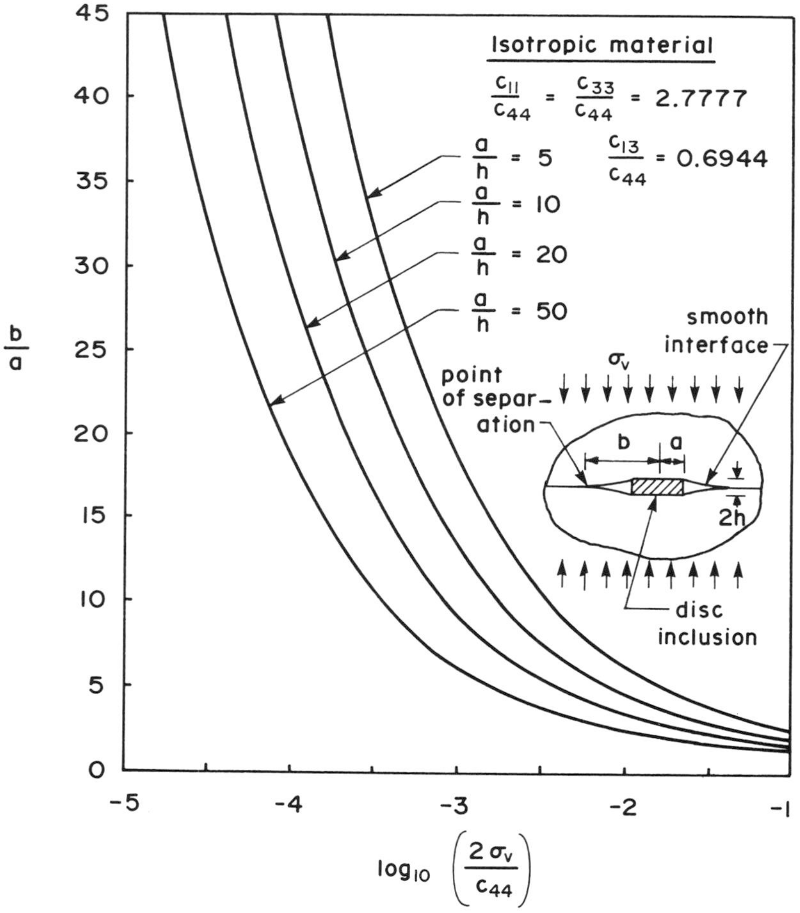

Figure 5: Influence of the precompression on the radius of the zone of separation - isotropic material

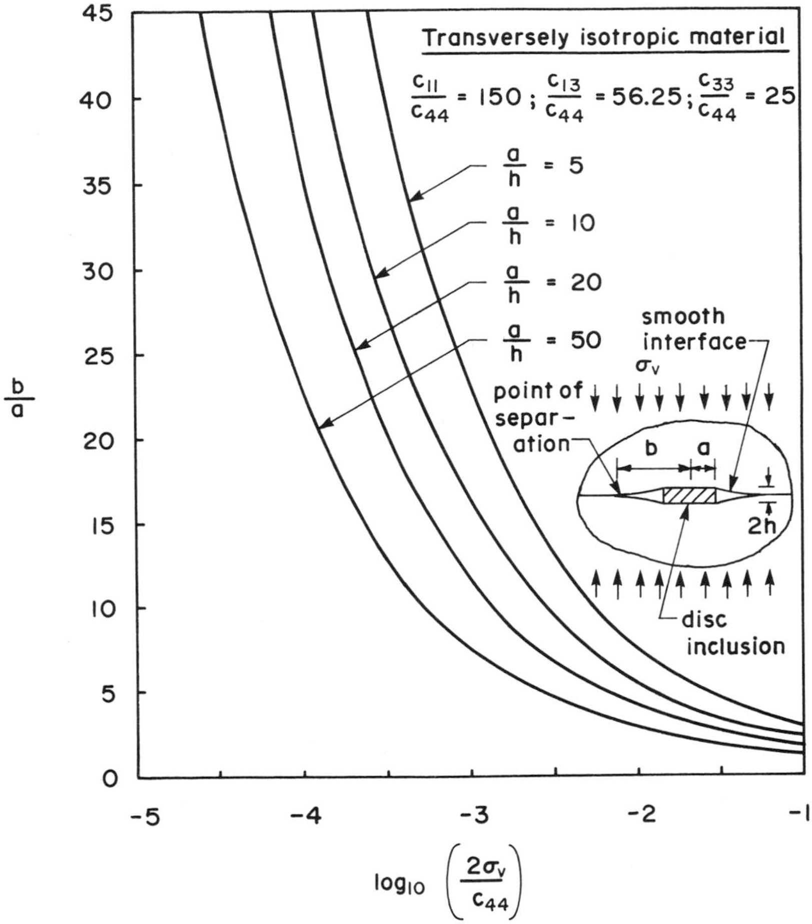

Figure 6: Influence of the precompression on the radius of the zone of separation - transversely isotropic material

REFERENCES

1 J. Dundurs and M. Stippes, J. Appl. Mech., 92 (1970) 965-970.
2 A.D. de Pater and J.J. Kalker, (Eds.), The Mechanics of Contact Between Deformable Media, Proc. IUTAM Symposium, Enschede, Delft University Press, Delft (1975) 414 pp.
3 G. Duvant and J.L. Lions, Inequalities in Mechanics and Physics, Springer Verlag, Berlin (1976) 397 pp.
4 G.M.L. Gladwell, Contact Problems in the Classical Theory of Elasticity, Sijthoff and Noordhoff, The Netherlands (1980) 716 pp.
5 L.M. Keer and K. Chantaramungkorn, J. Elasticity, 2 (1972) 191-197.
6 L.M. Keer, J. Dundurs and K.C. Tsai, J. Appl. Mech., 39 (1972) 1115-1120.
7 K.C. Tsai, J. Dundurs and L.M. Keer, J. Appl. Mech., 46 (1979) 198-202.
8 S.L. Pu and M.A. Hussain, J. Appl. Mech., 37 (1970) 859-861.
9 M.B. Civelek and F. Erdogan, J. Appl. Mech., 42 (1975) 136-140.
10 M.R. Gecit and F. Erdogan, Int. J. Solids Struct.,14 (1978) 771-785.
11 L.M. Keer and M.A.G. Silva, J. Appl. Mech., 39 (1972) 1121-1124.
12 M.B. Civelek and F. Erdogan, J. Appl. Mech., 43 (1976) 175-177.
13 M.R. Gecit, Int. J. Solids Struct., 16 (1980) 387-396.
14 M. Comninou, D. Schmueser and J. Dundurs, Int. J. Engng. Sci., 18 (1980) 131-137.
15 D. Schmueser, M. Comninou and J. Dundurs, Int. J. Engng. Sci., 18 (1980) 1149-1155.
16 D. Schmueser, M. Comninou and J. Dundurs, J. Engng. Mech. Div. Proc. ASCE, 107 (1981) 1103-1118.
17 A.P.S. Selvadurai and M.C. Au, Proc. 7th B.E.M. Conference (Ed. C. Brebbia and G. Maier) Como, Italy, Springer Verlag, Berlin, 2 (1985) 14.109-14.127.
18 A.P.S. Selvadurai, Trans. Canadian Soc. Mech. Engng., 7 (1983) 154-157.
19 G.M.L. Gladwell and T. Hara, Q.J. Mech. Appl. Math., 34 (1981) 251-263.
20 A.P.S. Selvadurai, J. Engng. Mech., Proc. ASCE, 110 (1984) 405-416.
21 J. Dundurs, K.C. Tsai and L.M. Keer, J. Elasticity, 3 (1973) 109-115.
22 Y. Weitsman, J. Appl. Mech., 36 (1969) 198-202.
23 G.M.L. Gladwell and K.R.P. Iyer, J. Elasticity, 4 (1974) 115-130.
24 G.M.L. Gladwell, in Mechanics of Contact Between Deformable Media, (Eds. A.D. de Pater and J.J. Kalker), Proc. IUTAM Symposium, Enschede, Delft Univ. Press (1975) 99-109.
25 O.J. Svec, Comp. Methods and Appl. Mech. Engng., 3 (1974) 105-113.
26 A.P.S. Selvadurai, Elastic Analysis of Soil-Foundation Interaction, Developments in Geotechnical Engineering, Vol. 17, Elsevier Scientific Publ. Co. Amsterdam (1979) 543 pp.
27 P.D. Panagiotopoulos, Inequality Problems in Mechanics and Applications. Birkhauser Verlag, Basel (1985) 432 pp.
28 M.P. Cleary, Proc. 20th U.S. Symp. Rock Mechanics, Univ. of Texas, Austin (1979) 127-142.
29 H.A. Elliott, Proc. Camb. Phil. Soc., 44 (1948) 522-531.
30 H.A. Elliott, Proc. Camb. Phil. Soc., 45 (1949) 621-630.
31 S.G. Lekhnitskii, Theory of Elasticity of an Anistropic Elastic Body, Holden Day, San Francisco (1963) 404 pp.
32 A.E. Green and W. Zerna, Theoretical Elasticity, Clarendon Press, Oxford (1968) 457 pp.
33 R.A. Eubanks and E. Sternberg, J. Rat. Mech. Anal.,3 (1954) 89-101.
34 J.C. Cooke, Q.J. Mech. Appl. Math., 16 (1963) 193-203.
35 W.E. Williams, Proc. Edin. Math. Soc., Ser. 2,13 (1963) 317-323.
36 C.J. Tranter, Proc. Glasgow Math. Assoc., 4 (1960) 200-203.
37 I.N. Sneddon, Mixed Boundary Value Problems in Potential Theory, North Holland, Amsterdam (1966) 283 pp.
38 R.P. Kanwal, Linear Integral Equations: Theory and Technique, Academic Press, New York (1971) 296 pp.
39 A.P.S. Selvadurai and B.M. Singh, Int. J. Fracture,25 (1984) 69-77.
40 K. Atkinson, A Survey of Numerical Methods for the Solution of Fredholm Integral Equations of the Second-Kind, Soc. for Industrial and Appl. Math., Phil. Pa (1976) 230 pp.

41 C.T.H. Baker, The Numerical Treatment of Integral Equations, Clarendon Press, Oxford (1977) 1034 pp.
42 A.P.S. Selvadurai (to be published).
43 B.I. Smetanin, Prikl. Math. Mech (PMM), 32 (1968) 458-462.
44 L.W. Moss and A.S. Kobayashi, Int. J. Fracture, 7 (1971) 89-99.
45 T. Shibuya, I. Nakahara and T. Koizumi, J. Appl. Math. Mech (ZAMM), 55 (1975) 395-402.
46 I. Choi and R.T. Shield, Int. J. Solids Struct., 18 (1982) 479-486.
47 A.P.S. Selvadurai and B.M. Singh, Q.J. Mech. Appl. Math., 38 (1985) 233-243.

LARGE ICE MASS SURGING VIA ICE-BEDROCK INTERFACE MOBILIZATION

D.F.E. Stolle[1] and F.A. Mirza[2]

[1] Assistant Professor, Dept. of Civil Engineering and Engineering Mechanics, McMaster University, Hamilton, Canada L8S 4L7

[2] Associate Professor, Dept. of Civil Engineering and Engineering Mechanics, McMaster University, Hamilton, Canada L8S 4L7

ABSTRACT

A finite element model for surging (instability) of large ice masses is presented. At the interface between the ice and the rock, the basal shear resistance is reduced according to the excess sliding energy dissipated above a certain threshold value. In addition to the threshold energy, the time for surging is also controlled by a lubrication factor. It is confirmed that a geothermal flux approaching 1.9 HFU is required to bring most of the south-west ice-bedrock interface of Barnes Ice Cap (Baffin Island, Canada) to pressure melting which would then allow for basal sliding and instability. It is also shown that the temperature changes during the period of surge of a polythermal ice cap are negligible.

INTRODUCTION

Unlike the apparently stable behaviour of most ice masses, a small group of glaciers undergo a rapid change in velocity and geometry within a short period of time after having been stable for a large number of years. These periods of fast flow or instability are referred to as catastrophic advances or surges (ref. 1). Very few surging glaciers have been well documented in the literature and therefore little is known about the mechanical properties and thermal regimes of these ice masses prior to surging. The lack of data for these ice masses has been attributed to the relatively rare occurrence of such events, the remoteness of their locations, and the inaccessibility of the glacier surface during instability (ref. 1).

There are two important aspects which should be considered in the study of instability of large ice masses: the conditions necessary for the attainment of high velocities; and how these velocities are maintained while the driving mechanism continuously decreases as the surge progresses, i.e. the ice mass flattens (ref. 2). Theoretical explanations based on mechanical or boundary instabilities, have been developed of which some include: nonunique longitudinal ice thickness profiles (ref. 3); basal temperature variations with time (refs. 3 and 4); and change in flow properties due to dynamic recrystallization (ref. 5). These explanations addressed possible trigger mechanisms but do not necessarily satisfy the conditions for surge propagation. Weertman (refs. 6 and 7) advanced a theory based on instability at the ice-bedrock interface due to formation of a thick water layer. Such instability can account for the high velocities observed for surging glaciers. Lliboutry (ref. 8) suggested that the basal instability can also result from excessive cavitation at the ice-bedrock interface at critical sliding velocities.

A large displacement, non-linear creep (implicit time-marching), plane strain finite element model with transient thermal analysis capability for conduction and convection has been developed to study surge propagation with the intent of investigating changes in velocity, stress and temperature fields during instability. To allow for basal sliding and sliding instability, a time-dependent sliding element has been developed with sliding resistance dependent on the energy dissipation at the ice-bedrock interface. It is demonstrated that a suitable choice of lubrication and threshold energy parameters can predict surge propagation of a large ice mass.

FINITE ELEMENT MODELS

It is not intended to review the development of the finite element models here, but rather to describe the creep and thermal models used for the surge propagation study. Owing to the short periods involved in surge propagations, the influence of mass balance can be assumed negligible. Thus, an updated Lagrangian description of deformation is convenient for studying the instability of large ice masses. The finite element equivalents for the stress equilibrium and the heat balance can be written as

$$[k_c]\ (\{\delta_{n+1}\} - \{\delta_n\}) - \{\Delta\bar{R}_n\} = \{\emptyset\} \tag{1}$$

$$([\bar{C}_n]/\Delta t_n+\theta[\bar{H}_n])\ \{T_{n+1}\}+(-[\bar{C}_n]/\Delta t_n+(1-\theta)[\bar{H}_n])\ \{T_n\}+\{Q_n\} = \{\emptyset\} \tag{2}$$

where $[k_c]$ depends on the implicitness parameter θ, creep and elastic stiffness; $[\bar{C}_n]$ is the heat capacitance matrix; $[\bar{H}_n]$ is the heat conductance matrix; $\{\Delta\bar{R}_n\}$ contains the creep loading; $\{Q_n\}$ is made up of the heating due to strains and the boundary fluxes; and $\{T_{n+1}\}$ and $\{T_n\}$ are the temperature vectors at times t_{n+1} and t_n, respectively. Similarly, $\{\delta_{n+1}\}$ and $\{\delta_n\}$ are the displacements at times t_{n+1} and t_n, respectively.

The stepwise uncoupling of the creep and thermal analyses is employed to complete the transient thermal creep analysis. It is assumed that the material response is isotropic and that the ice is homogeneous. Of course, variations in material properties can be easily handled. The geometry is immediately updated after each iteration to follow the large deformations that can occur during surges. In order to accommodate sliding at the ice-bedrock interface, a time-dependent sliding element has been developed. It is assumed that the basal shear stress $\tilde{\sigma}_1$ is related to the basal sliding velocity $\tilde{v}_1$ by

$$\tilde{\sigma}_1 = \tilde{k}^*_{11}(\tilde{v}_1, t_n)\ \tilde{v}_1 \qquad (3)$$

where $\tilde{k}^*_{11}(\tilde{v}_1, t_n)$ is a velocity- and time-dependent sliding parameter. It is by allowing $\tilde{k}^*_{11}(\tilde{v}_1, t_n)$ to vary with time that the basal instability is modelled. The tilde above the symbols refers to the local coordinate system along the mean bed slope.

Triangular elements have been employed for the transient models with linear displacements and temperature within an element. The decision to use these low order elements was based on the fact that the ice properties used can have reasonably large errors in their estimations thus undermining accuracies of the high performance elements.

SLIDING BOUNDARY ELEMENT

To accommodate sliding at the ice-bedrock interface, joint element with linear displacements due to Goodman et al. (ref. 9) has been extended to include the time-dependent sliding behaviour. Assumptions regarding the development of the modified joint element are:

(1) movement at the ice-bedrock interface is parallel to the mean bed slope;

(2) only the basal shear stress governs the sliding velocity at the ice-bedrock interface;

(3) the "soft" ice layer at the base also has some elastic resistance; and

(4) the soft ice layer is of negligible thickness.

The relationship between the basal shear stress $\tilde{\sigma}_1$ and the basal sliding velocity $\tilde{v}_1$ in Eqn. (3) is used and has been incorporated into incremental analysis in the following manner:

$$\{\Delta\tilde{w}\}_n = \{\Delta\tilde{w}^E\}_n + \{\Delta\tilde{w}^S\}_n = [N]\ \{\tilde{\delta}_{n+1} - \tilde{\delta}_n\} \tag{4}$$

where [N] contains the linear interpolation functions. Vectors $\{\Delta\tilde{w}\}_n$, $\{\Delta\tilde{w}^E\}_n$ and $\{\Delta\tilde{w}^S\}_n$ are the total, elastic and inelastic displacement increments of the ice, respectively, parallel and perpendicular to the bedrock.

The changes in stresses are related to changes in the elastic displacement increments as

$$\begin{Bmatrix} \Delta\tilde{\sigma}_1 \\ \Delta\tilde{\sigma}_2 \end{Bmatrix} = \begin{bmatrix} \tilde{K}_{11} & \emptyset \\ \emptyset & \tilde{K}_{22} \end{bmatrix}_n \begin{Bmatrix} \Delta\tilde{w}_1 - \Delta t_n \tilde{v}_1 \\ \Delta\tilde{w}_2 \end{Bmatrix}_n \tag{5}$$

$$\tilde{K}_{11} = \tilde{K}^*_{11}/(1 + \theta\Delta t_n \tilde{K}^*_{11}\ \partial\tilde{v}_1/\partial\tilde{\sigma}_1)_n \tag{6}$$

where $\tilde{K}_{11}$ and $\tilde{K}_{22}$ are the elastic spring constants tangent and normal to the bedrock, respectively; $\tilde{v}_1$ is given by Eqn. (3). The subscript n denotes time at t_n. Equation (5) introduces non-recoverable sliding through an initial displacement vector analogous to the initial strains in implicit creep analysis. The elastic constants $\tilde{K}_{11}$ and $\tilde{K}_{22}$ can be arbitrary for surging and a value of $1\emptyset^4$ kN/m^3 has been taken for the numerical applications to be presented.

The incremental finite element equations for equilibrium, using Eqn. (5) and the virtual work, are

$$[K_s]_n\ \{\tilde{\delta}_{n+1} - \tilde{\delta}_n\} - \{\Delta R\}_n = \{\emptyset\} \tag{7}$$

$$[K_s]_n = \int_{\Gamma_s} [N]^T\ [\tilde{K}]_n\ [N] d\Gamma \tag{8}$$

$$\{\Delta R\}_n = \Delta t_n \int_{\Gamma_s} [N_n]^T\ [\tilde{K}]_n\ \{\tilde{v}\}_n\ d\Gamma \tag{9}$$

where $[\tilde{K}]_n$ is the constitutive matrix of Eqn. (5), Γ_s is the portion of the boundary that slides and $\{\tilde{v}\}_n$ is the sliding velocity vector. It should be noted that the normal velocity $\tilde{v}_{2n}$ is assumed to be zero. Appropriate transformations are required to obtain $[K_s]_n$, in Eqn. (7), in the global coordinate system.

The time-dependent sliding boundary element presented above has been kept compatible with the three node triangular element for the creep and the transient thermal analysis. It also employs the same algorithm for the incremental creep analysis (ref. 10), using the implicit approach and the Newton-Raphson method as used for the triangular elements for the interior domain.

LARGE ICE MASS INSTABILITY

Previous investigations on surging of large ice masses suggest that the basal instability appears to be the main mechanism by which a surge propagates. To implement such instability, it is assumed that a loss in sliding resistance is due to formation and buildup of a water layer. The bedrock and ice-bedrock interface of ice masses likely have some permeability (ref. 11) so that a critical amount of energy must be dissipated to melt enough water to ensure that more water is produced locally than can run off. This excess water leads to the build-up of a water layer under a large ice mass and leads to subsequent instability.

To account for reduction in the basal shear resistance, it is suggested that the sliding parameter in Eqn. (3) be defined by:

$$\tilde{k}^*_{11} = k' + k^*/(1 + \langle \sum_i \phi_i (\tilde{\sigma}_1 \tilde{v}_1 - q_\emptyset)_i \, \Delta t_i \rangle) \tag{10}$$

where k^* is the initial shear resistance when no water has collected at the base; $\tilde{\sigma}_1 \tilde{v}_1$ is the local energy dissipation rate per unit area; $q_\emptyset$ is the threshold energy dissipation rate per unit area which accounts for the basal permeability; k' is a lower limit on the sliding parameter; Δt_i is the time increment during the ith creep iteration; and ϕ_i is a lubrication factor.

The reduction of shear resistance due to the energy dissipated in Eqn. (10) is somewhat similar to that proposed by Budd (ref. 12). In the present work, the influence of shear resistance reduction is introduced through a constitutive relationship rather than directly through the stresses as was done by Budd. It is implied that the summation within the brackets < > cannot be less than zero. The lubrication and threshold energy dissipation parameters, both influence the surge timing. These parameters can change with position under the ice mass and with time due to changes at the ice-bed interface as a surge propagates.

In the present study, $q_{\emptyset}$ was taken as zero, i.e. all numerical simulations started at near ripe conditions for instability for economizing on computational cost. The minimum value for $\tilde{k}^{*}_{11}$ was chosen as Ø.Ø1 $kN{\cdot}yr/cm^3$. The lubrication factor was estimated, through trial runs, to be between Ø.Ø5 and 1.Ø m/kN for a surge lasting from one to two years. The value of ϕ_i depends upon the initial shear resistance k*, which is a function of the bedrock topography and the thermal conditions. The instability model adopted in this study was chosen for its simplicity and, at the same time, its ability to model a reduction in the basal shear resistance according to the events that occur at the ice-bedrock interface. Of course, other available sliding models can be easily incorporated into the finite element algorithm.

IDEALIZED DOUBLE SLOPE ICE MASS

The idealized double slope ice mass, shown in Fig. 1, was analyzed to illustrate the influence of the lubrication factor ϕ_i on the timing of a surge. Hooke's power law parameters (see Eqn. (11)) were used to model the creep behaviour; parameters $k' = \emptyset$ and $k^* = 1\emptyset\emptyset\emptyset$ were employed for the sliding model in Eqn. (1Ø). Figures 2(a) and 2(b) show the changes in the horizontal surface velocity at node 25 and in stresses in element 19, respectively, due to reducing ϕ_i. As anticipated, horizontal velocities increased due to an increase in equivalent stress σ_e (Dorn's definition) as a result of decreased basal shear resistance. The increase in σ_e was a result of increasing longitudinal strains due to decreasing shear stress σ_{12}. This example demonstrates clearly that the timing of a surge for a large ice mass can be controlled by the lubrication factor.

SURGE SIMULATIONS OF BARNES ICE CAP

A finite element simulation was completed along section N-O-SS, shown in Fig. 3, of Barnes Ice Cap due to suitability of plane strain flow assumptions along this section. Furthermore, ample evidence exists that supports the hypothesis of instability having taken place on the south-west side of the ice cap (see ref. 13), and a reasonable amount of information on the thermal regime of the ice cap is also available from which some conclusions can be

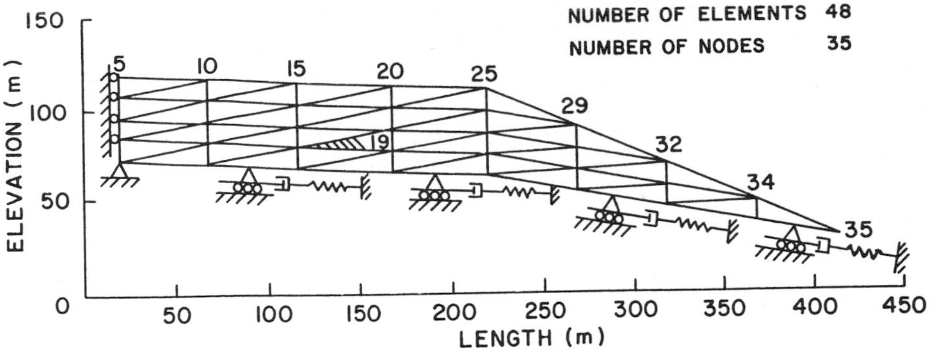

Fig. 1. Finite element mesh and boundary conditions for the double slope ice mass.

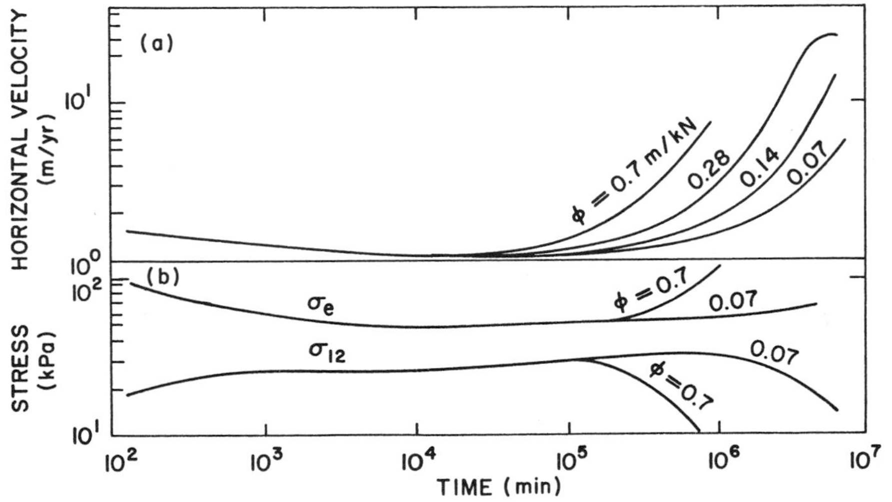

Fig. 2. Plots of (a) horizontal surface velocity at node 25 and (b) stresses in element 19 versus time for different values of the lubrication factor ϕ.

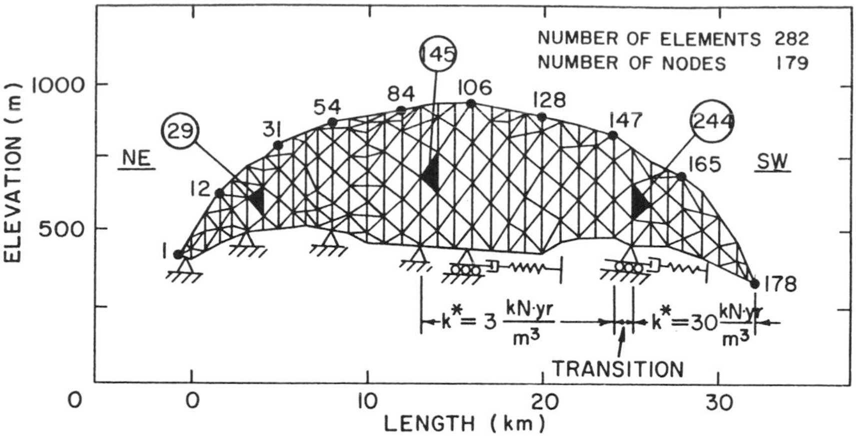

Fig. 3. Finite element mesh and boundary conditions for Barnes Ice Cap.

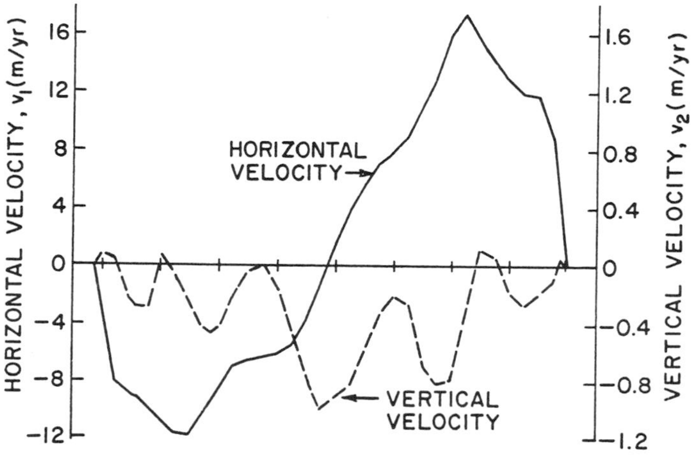

Fig. 4. Horizontal and vertical velocities at the surface of Barnes Ice Cap at near steady-state conditions.

drawn regarding mechanism for the surge propagation.

Classen (ref. 14), by thermal drilling and deep ice-temperature measurements along the flow line of interest, confirmed that the temperate basal ice exists near the ice-bedrock interface. This temperate basal ice on the south-west side is favourable for basal instability. According to Holdsworth (ref. 15), instability may be attributed to an outward spreading of this temperate basal ice. This reduces the passive resistance which prevents the advancement of the ice cap margin. Once the contact area between the ice frozen to the bedrock reduces to a critical size, as determined by the ice strength properties, the basal ice yields or flows and a surge propagates. Field studies show the existence of recessional morraines which indicate that the ice cap has been receding for several centuries (ref. 16). The computer simulations by Mahaffy (ref. 17) also support the hypothesis that the south dome is retreating. Based on this information it is felt that the onset of instability may have been partially caused by a loss in passive resistance due to the receding margin. In order to model the surge initiation, factors influencing the long-term ice cap dynamics and energy balance must also be incorporated into the model. However, this is left for future improvements.

The emphasis in the current study is placed on modelling surge propagation on the south-west side of Barnes Ice Cap via basal instability and by taking into account the interaction between the temperature and velocity fields during the surge. The pre-surge ice mass profile has been provided by Holdsworth (ref. 18) of Environment Canada. The finite element idealization for the pre-surge profile of Barnes Ice Cap is shown in Fig. 3 and the following data were adopted.

(1) Flow law (Hooke's Parameters)

$$\dot{\varepsilon}_e^c = 0.0327\ \sigma_e^{1.65} \tag{11}$$

where $\dot{\varepsilon}_e^c$ (yr^{-1}) and σ_e (0.1 MPa) are according to Dorn's definitions for equivalent strain and stress, respectively.

(2) Thermal conductivity $k_T = 2.1\ W \cdot m^{-1} \cdot {}^oK$

(3) Heat Capacity $C = 2.1 \times 10^{-6}\ W \cdot s \cdot m^{-3} \cdot {}^oK$

(4) Flow law including temperature dependence

$$\dot{\varepsilon}_e = f(\sigma_e)\ \exp(-Q/RT) \tag{12}$$

where $f(\sigma_e)$ has the same form as Eqn. (11) and the constant 0.0327 corresponds to $265^\circ K$; R is the gas constant; and

$Q = 78$ kJ/mol for $T \geq 265^\circ K$

$Q = 120$ kJ/mol for $T < 265^\circ K$.

The study on instability is divided into two parts: preliminary investigation to establish initial conditions based on steady-state analysis; and surge simulation with starting conditions that are favourable for a surge. Instability is assumed to be due to a local reduction of basal shear resistance according to the model that has been presented previously.

Preliminary investigation

To perform the non-isothermal surge simulation an initial temperature distribution is required. Since no data is available on the pre-surge temperature field of BIC, it is assumed that the ice cap adjusted to a near steady-state temperature field before the surge (ref. 18). In order to estimate a velocity field for the steady-state thermal analysis, an isothermal creep simulation was completed where no sliding was allowed at the ice-bedrock interface. The surface velocities from this analysis are shown in Fig. 4.

The steady-state temperature simulations were completed by using the velocity, strain heating and pressure fields from the isothermal creep analysis. Although the computed temperature and creep responses are not strictly compatible, it was believed that the computed temperature field would be reasonable in view of the lack of information on the pre-surge conditions. One could perform interactive creep and thermal analyses to arrive at more compatible velocity-thermal fields. However, such a refined analysis would be meaningful only if very accurate data were available regarding physical properties of ice, geometry and boundary conditions.

The boundary conditions for the steady-state temperature simulation are shown in Fig. 5. Four steady-state thermal simulations were completed for different geothermal flux values ranging from 0.95 HFU to 1.9 HFU. It was observed that for an appreciable portion of the base to reach the pressure melting state, the geothermal flux must be sufficiently high, i.e. 87.5 percent of 1.9 HFU. This observation is consistent with the hypothesis that the geothermal flux is higher on the south-west side than on the north-east side of the current divide (ref. 18). The letters at the base of the ice cap indicate geothermal flux levels required for the ice to attain the pressure melting state. Fig. 5 also shows the temperature distribution over the ice cap that was used for the non-isothermal surge simulation to follow.

Since the south-west margin of Barnes Ice Cap is frozen to the bedrock as indicated by the temperature contours in Fig. 5, the basal ice at the margin must yield for a surge to propagate. A separate elasto-visco-plastic analysis confirmed that the basal ice would yield at the margin. However, to simplify the surge analyses, no distinction was made between the loss in shear resistance due to lubrication by the melt water and ice yielding.

Surge simulation

The sliding parameters for surging are very difficult to determine. Therefore, it was necessary to make assumptions for these parameters. Preliminary investigation showed that an initial sliding resistance, $k^* = 3$ kN • yr • m^{-3}, was found to predict reasonable horizontal velocities when compared with measured velocities for the post-surge simulations of Barnes Ice Cap. A larger initial shear resistance toward the south-west margin, $k^* = 30$ kN • yr • m^{-3}, was used to reflect the anticipated higher resistance during the early stages of the surge. It was also found that a lubrication factor $\phi_i = 0.05$ m/kN would ensure near completion of the surge within two years for the range of initial sliding resistances above. Neither the sliding parameters nor their distributions have been optimized and has been left for a future parametric study. The boundary and the initial conditions for the surge simulation are shown in Figs. 3 and 5.

During some preliminary surge simulations, it was found that advancements of the south-west margin during a surge are distinctly influenced by the lubrication factor, initial basal shear resistance and the temperature field. The influence of the temperature field on sliding enters through coupling between the internal deformation (creep) and stresses in the basal layer.

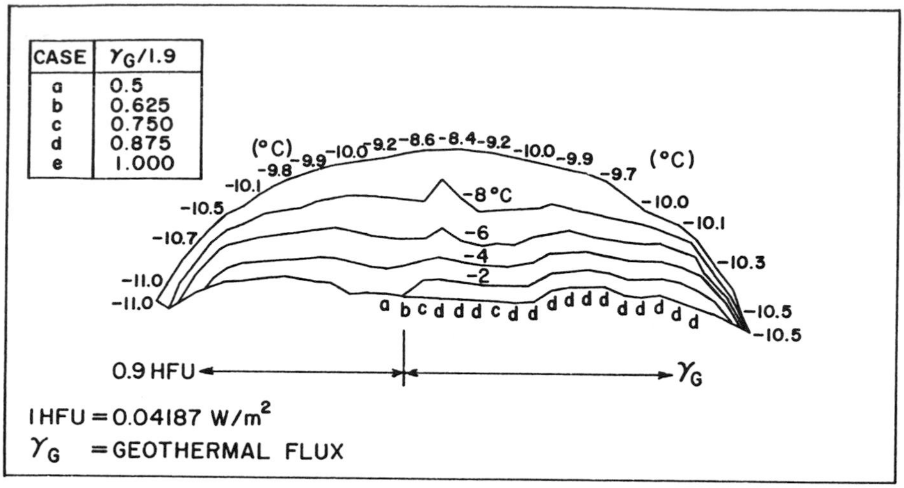

Fig. 5. Boundary conditions for thermal analysis of Barnes Ice Cap and computed temperature contours.

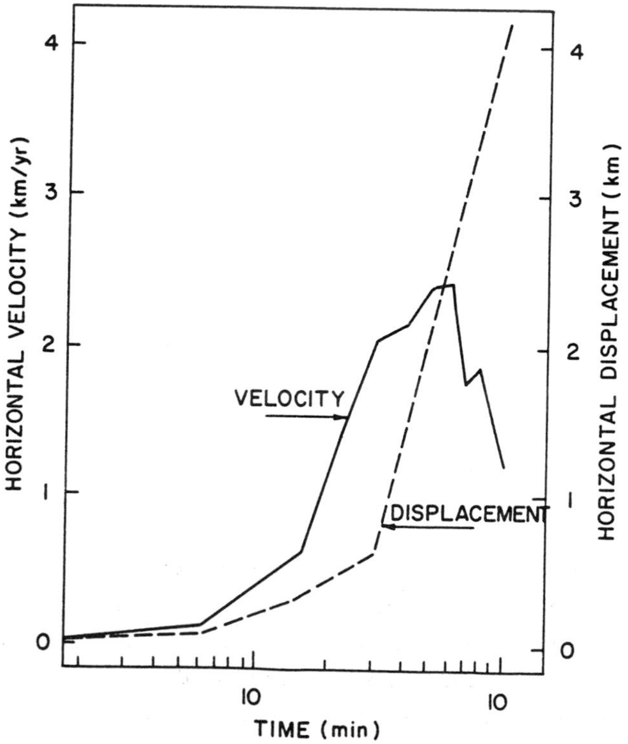

Fig. 6. Time variation of horizontal velocity and displacement at node 165 (Fig. 3).

Fig. 6 shows the advancement of the south-west toe and changes in the horizontal velocity at node 165 with time for the set of parameters used. It was observed that after 1.9 years of surging, the toe advanced more than 4 km; most of the advance occurred after the first year. The horizontal velocity at node 165 increased, slowly in the earlys stages, and then started to decrease. This decrease in velocity was due to spreading and lowering of the ice surface and an increase in the ice-bedrock contact area. The ice margin advanced 1.7 km when the velocity reached its highest value of 2.4 km/yr.

The changes in equivalent stresses at various locations are shown in Fig. 7. The equivalent stress changes in element 29 are indicative of the transient behaviour near the north-east margin, during the surge. It appears that the equivalent stress in at least some of the north-east margin increased slowly due to surging on the south-west side. Although this increase is much smaller than the increase in equivalent stress in the surge zone, as expected. A comparison of the equivalent stresses in elements 145 and 244 indicates that the surge started to propagate first in the middle of the ice mass and then spread toward the margin. This is because the initial basal shear resistance at the margin was much greater than that near the middle. The increase in equivalent stress during surging reflected the increase in the longitudinal extension as a result of the loss in the basal shear resistance. The eventual decrease of equivalent stress suggested that the ice cap was trying to stabilize. This was also reflected by the changes in the velocity field as indicated previously.

During the surge, the basal shear stress was observed to decrease as shown in Fig. 8. An increase in the peak shear stress could have been achieved by reducing the distance over which the initial sliding resistance k^* was equal to 30.0 kN · yr · m^{-3}.

The profiles of Barnes Ice Cap, before the surge started and near the completion of the surge (approximately 2 years), are shown in Fig. 9. This figure demonstrates how the surface of the ice cap lowered and flattened as the margin advanced. It can be observed that as the surge propagated, the location of the dynamic divide, which was originally near the 16 km section, approached the position of the current divide (near the 10 km section). It should be noted that the shape of Barnes Ice Cap immediately after the surge as predicted by the finite element model was reasonable if one took into account the melting of the

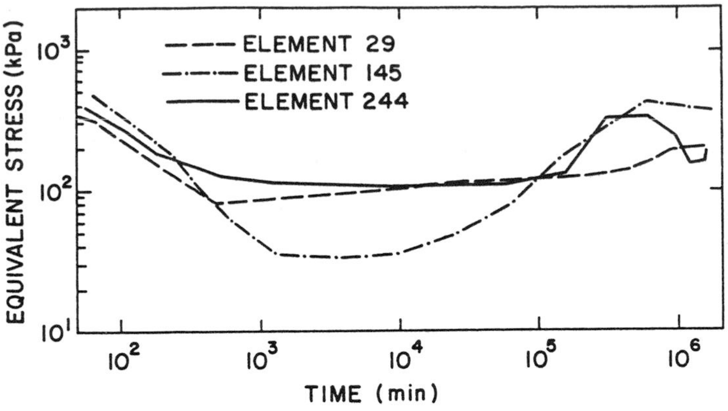

Fig. 7. Time variation of equivalent stress σ_e in various elements (Fig. 3).

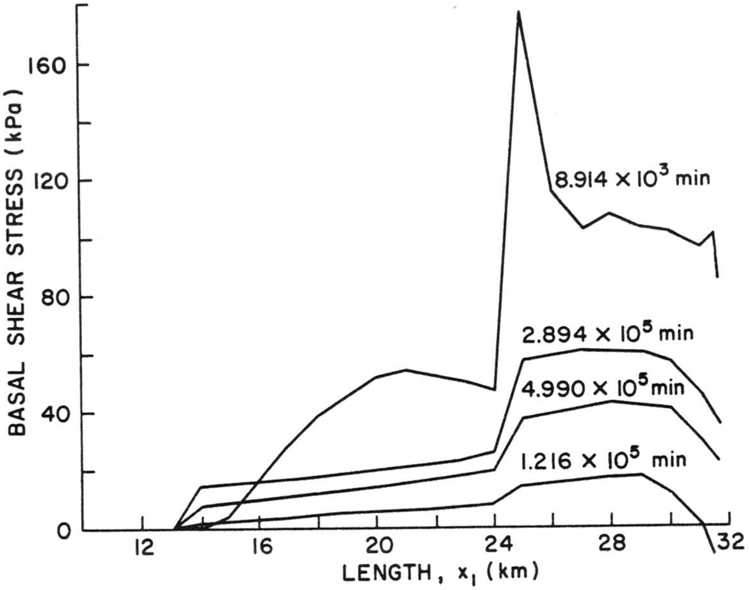

Fig. 8. Basal shear stress distribution at various instants of time.

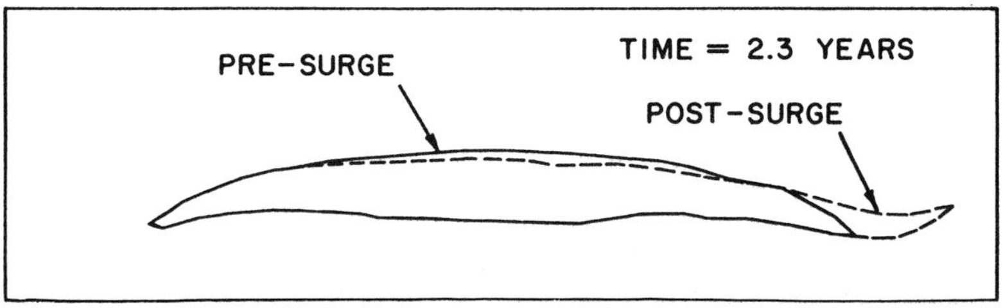

Fig. 9. Pre- and post-surge profiles of Barnes Ice Cap.

margin and the long-term creep of the ice cap.

Conservative estimates for the temperature increases at the ice-bedrock interface indicated that these changes should be limited to 0.07°C at most. The temperature increases, as predicted by the model, are well within this estimate. The post-surge contour plot of the temperature field was essentially the same as shown in Fig. 5. Thus it was found that spreading of the temperate ice zone is minimal due to the small temperature changes. This observation is in agreement with that of Clarke (ref. 19) who found that a thick layer of temperate basal ice does not form as a result of a surge. This is obviously contrary to Robin's (ref. 3) suggestion that a thick temperate basal ice layer forms during an instability.

SUMMARY AND CONCLUSIONS

For the Barnes Ice Cap, pre-surge temperature simulations indicate that a geothermal flux approaching 1.9 HFU is required to bring most of the south-west ice-bed interface to pressure-melting. The frozen margin that initially blocks a surge must yield plastically. Not enough frictional heat is dissipated to bring the ice near the margin to pressure melting. The initial conditions are very important for modelling the instability. These conditions include the temperature distribution within the ice mass and the spatial variation of basal shear resistance. Although the following influences have not been considered, the flow relationship, non-homogeneity and non-isotropy of the ice mass may be important and need further investigation. The basal shear resistance model used in this study has been adopted for its simplicity and its ability to accommodate a time-dependent loss of the basal shear resistance.

This study demonstrates that surge simulations, which include the influences of temperature and basal instability, can be modelled by using the finite element method. For the choice of parameters selected in this study, the results obtained seem very reasonable. It is expected that a better choice of these parameters, determined on the basis of parameteric study should yield better results. A logical next step is to also include the surge initiation as a part of the finite element modelling of large ice mass instability.

REFERENCES

1 W.S.B. Paterson, The Physics of Glaciers, Pergamon Press, London, 1969.

2 M.F. Meier, Seminar on the causes and mechanics of glacier surges, St. Hilaire, Canada, Sept. 10-11, 1968: A summary. Can. J. Earth Sci., Vol. 6, No. 4, Pt. 2 (1969), 987-89.

3 G. de Q. Robin, Initiation of glacier surges, Can. J. Earth Sci., Vol. 6, No. 4, Pt. 2 (1969), 919-27.

4 G. de Q. Robin, Ice movements and temperature distributions in glaciers and ice sheets, J. Glaciology, Vol. 2, No. 18 (1955), 523-32.

5 J.J. Jonas and F. Müller, Deformation of ice under high internal shear stresses, Can. J. Earth Sci., Vol. 6, No. 4, Pt. 2 (1969), 963-967.

6 J. Weertman, Catastrophic glacier advances. Union Géodesique et Géophysique Internationale. Association International d'Hydrologie Scientifique. Commission des Neiges et des Glaces. Colloque d'Obergurgl, Sept. 10-18, 1962, p. 31-39.

7 J. Weertman, The unsolved glacier sliding problem, J. Glaciology, Vol. 23, No. 89 (1979), 97-115.

8 L.A. Lliboutry, General theory of subglacial cavitation and sliding of temperate glaciers, J. Glaciology, Vol. 7, No. 49 (1968), 21-58.

9 Goodman, R.E., Taylor, R.L. and Brekke, J.L., A model for the mechanics of jointed rock, Soil Mech., Found. Div., ASCE, Vol. 94, No. SM3 (1968), 637-659.

10 M.B. Kanchi, O.C. Zienkiewicz and D.R.J. Owen, The visco-plastic approach to problems of plasticity and creep involving geometric nonlinear effects, Int. J. Num. Methods Engg., Vol. 12 (1987), 169-181.

11 L.A. Lliboutry, Local friction laws for glaciers: A critical review and new openings, J. Glaciology, Vol. 23, No. 89 (1979), 67-95.

12 W.F. Budd, A first simple model for periodically self-surging glaciers, J. Glaciology, Vol. 14, No. 70 (1975), 3-21.

13 G. Holdsworth, Evidence of a surge on Barnes Ice Cap, Baffin Island, Can. J. Earth Sci., Vol. 10, No. 10 (1973), 1565-74.

14 D.F. Classen, Temperature profiles for Barnes Ice Cap, surge zone, J. Glaciology, Vol. 18, No. 80 (1977), 391-405.

15 G. Holdsworth, Surge activity on the Barnes Ice Cap, Nature, Vol. 269, No. 5629 (1977), 588-90.

16 O.H. Løken, and J.T. Andrews, Glaciology and chronology of fluctuations of the ice margin at the south end of the Barnes Ice Cap, Baffin Island, N.W.T., Geographical Bulletin, Vol. 8, No. 4 (1966), 341-59.

17 M.W. Mahaffy, A three-dimensional numerical model of ice sheets: Tests on the Barnes Ice Cap, N.W.T., J. Geophys. Res., Vol. 81, No. 6 (1976), 1055-66.

18 G. Holdsworth, Barnes Ice Cap thermal regime-outline, Private communication (1978).

19 G.K.C. Clarke, Thermal regulation of glacier surging, J. Glaciology, Vol. 16, No. 2 (1975), 231-59.

AN AXISYMMETRIC INTERFACE ELEMENT FOR SOIL-STRUCTURE INTERACTION PROBLEMS

M. O. FARUQUE
Department of Civil and Environmental Engineering, University of Rhode Island, Kingston, RI 02881 (USA)

ABSTRACT

An axisymmetric interface element is used to model interfaces between dissimilar materials. A parametric study is made to identify the variables affecting the rate of convergence. Two boundary value problems are analyzed using an axisymmetric finite element procedure developed by the author. Results are presented for different interface conditions and for various values of aspect ratios and stiffness ratios. It is found that the element used herein can adequately model the slip and nonslip shearing modes of interfaces.

INTRODUCTION

Problems in Geomechanics and Structural Mechanics often include interfaces between dissimilar materials. If throughout the loading history perfect bonding is maintained, then the presence of an interface offers no difficulties; the interface merely serves as a boundary between the two adjacent materials. However, if the bond breaks down and there is relative movement of the two mating surfaces, then special analysis procedures are required. Examples of problems in which material interfaces play a prominent role include soil-structure interaction problems i.e. soil-retaining wall, soil-footing, soil-pile system, soil-culvert system, underground structures etc. In reinforced concrete or reinforced earth, the bond between the reinforcement and the surrounding matrix material is important. Because if the bond breaks down and relative slip occcurs along the reinforcement, the system becomes nonlinear and inelastic. In general, rock mass includes discontinuity, joints and weak planes which need special attention during the analyses of problems involving such materials. Besides there are many classical contact problems where interface play an important role.

The stress-strain responses of interface materials are, in general, different from the adjacent continuum. This is due to the fact that an interface can experience slip and/or opening during a deformation process creating discontinuity in the medium. Moreover, due to the irreversible nature of the deformation, energy dissipation takes place along interfaces. Thus redistribution of the stresses are necessary to maintain overall equilibrium. This, however, requires special solution technique such as incremental or iterative schemes (ref.1).

The finite element method is a versatile numerical procedure for the analyses of many boundary value problems whose solutions are either difficult to obtain or intractable. Soil-structure interaction problems often fall into this category. This class of problems are further complicated by the presence of interfaces which are highly nonlinear and inelastic. The finite element method can be employed to analyze soil-structure interaction problems, provided special elements are used to model the interface region.

A large number of interface elements are currently available (ref.2-5, 7-10, 12, 13, 15). Details of these interface elements and their relative merits and demerits can be found elsewhere (ref.6, 16).

The objective of this paper is to present an axisymmetric interface element for modelling joints and interfaces in soil-structure interaction problems. The interface element has finite thickness which makes it particularly suitable for modelling rock joints, faults or seams. Rate of convergency of the iterative scheme is studied in details using two hypothetical problems. It is found that the convergence rate is significantly influenced by two parameters S_R and A_R where S_R denotes the relative stiffness of the interface compared to the adjacent elastic medium and A_R denotes the aspect ratio (length/thickness) of the interface element. As a final step, a number of boundary value problems are analyzed using an axisymmetric finite element procedure. Results are presented for different interface conditions which are of practical interest.

INTERFACE ELEMENT

Fig. 1 shows a typical four noded axisymmetric interface element. The thickness of the interface element is assumed to be small compared to its length.

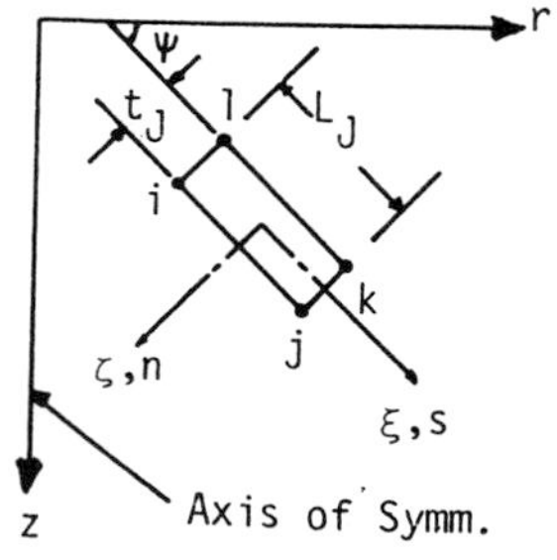

Fig. 1. Typical four-noded interface element

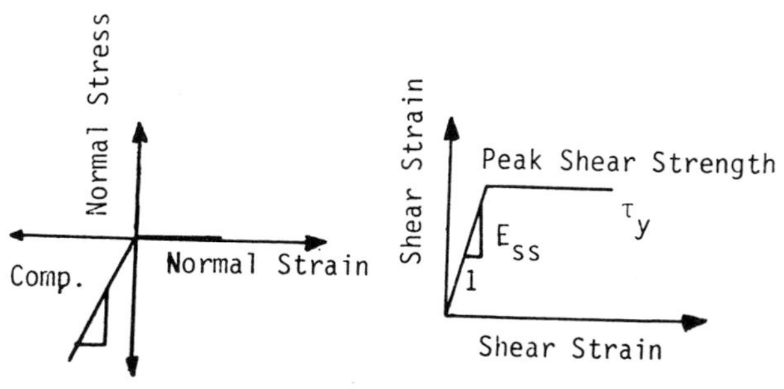

Fig. 2. Assumed stress-strain curves for the interface material

Thus both stresses and strains can be assumed to be constant across the thickness of the interface element. The interpolation functions along the length of the element is assumed to be

$$h_L = (1 - s)/2; \; h_R = (1 + s)/2 \tag{1}$$

where s is a nondimensional local coordinate along the length of the interface element (see Fig. 1) such that

$$-1 \leqslant s \leqslant 1 \tag{2}$$

Subscripts L and R in Eq. 1 denote the left and the right ends of the interface element respectively.

Using the interpolation function in Eq. 1, the relative displacements can be expressed as

$$\Delta u = h_L(u_l - u_i) + h_R(u_k - u_j) \tag{3}$$

$$\Delta w = h_L(w_l - w_i) + h_R(w_k - w_j) \tag{4}$$

where Δu and Δw are the relative radial and axial displacements respectively. Similarly the radial displacement along the bottom of the interface element can be written as

$$u^B = h_L u_i + h_R u_j \tag{5}$$

The quantities u_i, u_j, u_k, u_l, w_i, w_j, w_k, w_l are the nodal displacements. The relative displacements Δu and Δw can be transformed into local values Δu_s and Δw_n by using the following equations:

$$\begin{bmatrix} \Delta u_s \\ \Delta w_n \end{bmatrix} = \begin{bmatrix} \cos\psi & \sin\psi \\ -\sin\psi & \cos\psi \end{bmatrix} \begin{bmatrix} \Delta u \\ \Delta w \end{bmatrix} \tag{6}$$

where Δu_s and Δw_n are the relative displacements along the local coordinates s and n respectively and ψ is the orientation of the interface element (Fig. 1). The strains can be defined as

$$\varepsilon_{ss} = \Delta u_s / t_J \tag{7}$$

$$\varepsilon_{nn} = \Delta w_n / t_J \tag{8}$$

$$\varepsilon_{\theta\theta} = (u^B + \Delta u/2)/r \tag{9}$$

where t_J is the thickness of the interface element, r is the radial coordinate. ε_{ss} and ε_{nn} are the shearing and normal strains respectively. $\varepsilon_{\theta\theta}$ is the circumferential strain.

The elastic constitutive matrix may be defined as

$$[E] = \begin{bmatrix} E_{ss} & 0 & 0 \\ 0 & E_{nn} & 0 \\ 0 & 0 & E_{\theta\theta} \end{bmatrix} \tag{10}$$

where E_{ss} is the shear modulus and E_{nn} is the normal modulus for the interface. $E_{\theta\theta}$ is assumed to be zero for the present analysis.

Using Eqs. 1 through 10, the stiffness matrix of the interface element can be obtained from the integral

$$[K_J]_e = \int_V [B]^T [E] [B] \, dV \tag{11}$$

where

$$[B] = \begin{bmatrix} -B_1 & -B_2 & -B_3 & -B_4 & B_3 & B_4 & B_1 & B_2 \\ B_2 & -B_1 & B_4 & -B_3 & -B_4 & B_3 & -B_2 & B_1 \\ B_5 & 0 & B_6 & 0 & B_6 & 0 & B_5 & 0 \end{bmatrix} \tag{12}$$

and

$$\begin{aligned} B_1 &= h_L \cos\psi / t_J & \qquad B_4 &= h_R \sin\psi / t_J \\ B_2 &= h_L \sin\psi / t_J & \qquad B_5 &= h_L / 2r \\ B_3 &= h_R \cos\psi / t_J & \qquad B_6 &= h_R / 2r \end{aligned} \tag{13}$$

The stiffness matrix given by Eq. 11 is a tangent stiffness matrix. Since the interface behavior is highly nonlinear and inelastic in nature, a simple elastic-perfectly plastic shear stress-strain relationship is assumed for the interface material. The peak shear strength is a material constant in this model which can be determined from a direct shear test. The normal stiffness in compression is assumed to be high in order to avoid interpenetration. During debonding a small value of the normal stiffness is assigned. Fig. 2 shows the elasto-plastic stress-strain curves utilized for the present analysis.

NONLINEAR SOLUTION TECHNIQUE

The initial stress approach (ref.15) is employed as the iterative scheme to satisfy equilibrium. In this method the stiffness of the system is kept constant while the load vector is modified at the commencement of each iteration step. Fig. 3 shows a schematic of this iterative scheme.

The iterative technique used herein is such that the total load is applied in one step and equilibrium is satisfied through successive iterations. Thus the

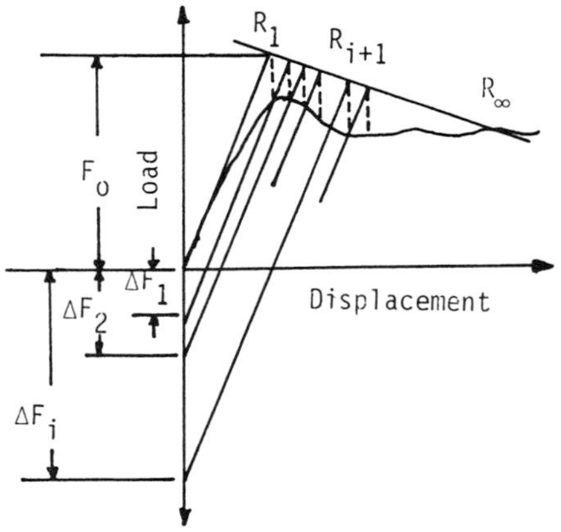

Fig. 3. Schematic of the initial stress approach

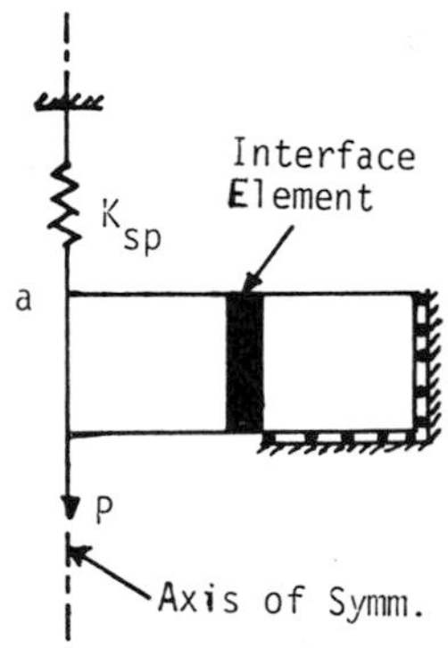

Fig. 4. Two rigid blocks with an interface element

solution yields only one point on the load-displacement curve. A history of deformation, however, can be obtained by solving the problem a number of times using different loads. An advantage of this technique is that it can yield a solution which falls on the strain softening side of the curve (see Fig. 3).

CONVERGENCY

To investigate the performance of the interface element and to study the rate of convergence, the foregoing interface element is implemented into an axisymmetric finite element procedure. Two hypothetical problems are then solved using the algorithm and influence of various parameters are investigated. These results are presented in the following sections:

Case 1

Fig. 4 shows two rigid blocks separated by an interface element. The outer rigid block is restrained from movement while the inner block is supported by an elastic spring of stiffness, K_{sp}. The system is axisymmetrically loaded by a

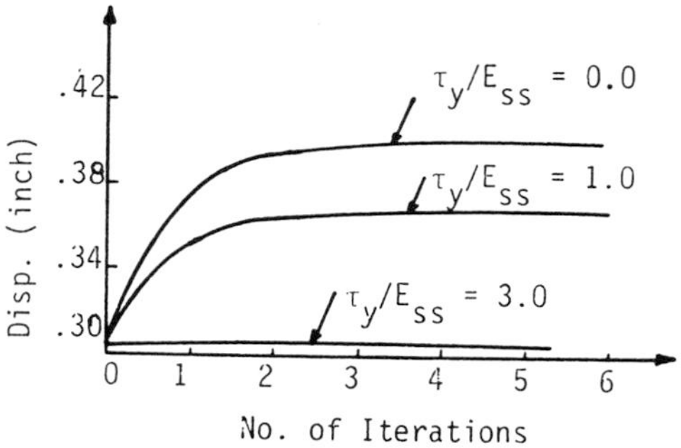

Fig. 5. Convergency of displacements for aspect ratio, A_R = 100

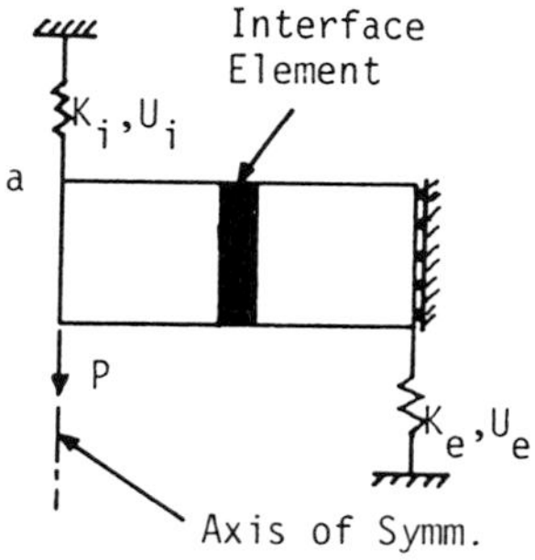

Fig. 6. Two rigid blocks separated by an interface element

single concentrated load, P. The variables which are important for the study of convergency are the following:

$S_R = K_{sp}/(E_{ss}/t_J)$	=	Ratio of the spring stiffness to the joint stiffness
$A_R = L_J/t_J$	=	Aspect ratio i.e. ratio of length of joint element to thickness
$T_R = \tau_y/E_{ss}$	=	Ratio of the limiting shear strength (τ_y) to the shear modulus E_{ss} of the interface element

Table 1. shows the influence of S_R, A_R and T_R on the rate of convergency of the iterative scheme. It is found that for $A_R>10^4$ the rate of convergency is extremely slow. It is also evident from Table 1. that the parameter T_R does not significantly influence the rate of convergence. To investigate the influence of the parameter, S_R, the same problem has been solved keeping S_R equal to constant (S_R=20). Results are presented in Table 2. It is seen that in this case even with $A_R = 10^5$, relatively small number of iterations are required for convergence. Thus both A_R and S_R are important for the rate of convergency of the iterative scheme.

Fig. 5 shows the schematic of the rate of convergence for A_R = 100 and S_R = 20 for various values of T_R. It is seen that for all the cases convergence is achieved within three iterations.

TABLE 1
Effect of A_R on the rate of convergency (K_{sp} = 100000)

Aspect ratio, A_R	T_R = 0.0	T_R = 1.0	T_R = 3.0	$\frac{K_{sp}}{E_{ss}/t_J}$
	No. of iter.	No. of iter.	No. of iter.	
10	3	0	0	200
100	7	6	4	20
1000	29	28	27	2
10000	182	179	172	0.2
100000	does not converge	-	-	0.02

TABLE 2
Effect of A_R and S_R on the rate of convergency (T_R = 0.0)

Aspect ratio, A_R	No. of Iteration	Spring ratio, S_R
10	6	20
100	7	20
1000	7	20
10000	10	20
100000	28	20

Case 2

Fig. 6 shows two rigid blocks separated by an interface element. The rigid blocks are axisymmetric and supported by springs of stiffnesses K_i and K_e where the subscripts i and e refers to the interior and exterior block respectively. The system is loaded axisymmetrically by a single concentrated load, P. This problem is intended to study the effect of the relative stiffnesses of the medium separated by the interface on the rate of convergency of the iterative technique. An aspect ratio, A_R, of 100 is used for this problem. Fig. 7 shows how the number of iterations required for convergency varies as a function of the stiffness ratio K_e/K_i. It is seen from Fig. 7 that when $K_e/K_i \simeq 0.1$, the system needs large number of iterations to converge. For $K_e/K_i > 0.1$ or $K_e/K_i < 0.1$, however, convergency is achieved with relatively few iterations. This demonstrates that the relative stiffnesses of the adjacent mediums separated by the interface is also an important parameter for the convergency of the iterative scheme.

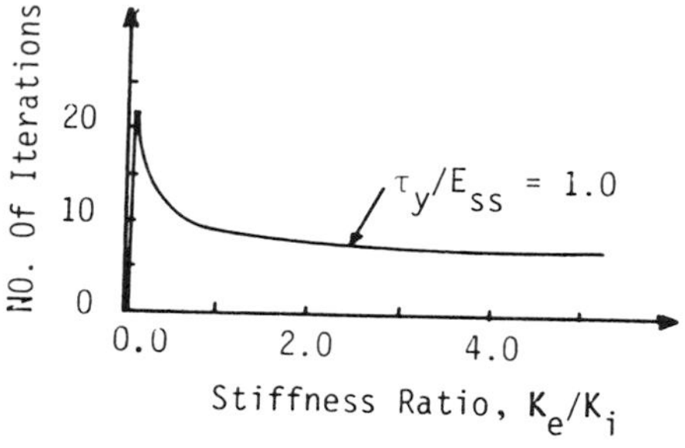

Fig. 7. Effect of stiffness ratio, S_R on the rate of convergency

BOUNDARY VALUE PROBLEMS

It is mentioned earlier that the interface conditions plays an important role in the analyses of many soil-structure interaction problems. Two such problems are investigated herein using an axisymmetric finite element procedure developed by the author. The continuums on either side of the interface are assumed to be homogeneous , isotropic and linearly elastic. A simple elastic-perfectly plastic shear stress shear strain response curve is used for the interface material. Results are presented in the following sections:

Circular plate resting on a halfspace

Circular foundations are used in many engineering structures such as silos, nuclear power plants, chimneys, towers, etc. The response of such foundations are dependent upon the interface condition between the plate and the soil.

Both rigid and flexible plates are analyzed here. The rigidity of the plate is expressed in terms of a parameter K_r (ref.6,14) defined by

$$K_r = \frac{\pi(1 - \nu^2)}{6(1 - \nu_p^2)} \frac{E_p}{E} \left(\frac{h}{a}\right)^3 \tag{14}$$

where E, ν are the Young's modulus and the poisson's ratio of the halfspace respectively. E_p and ν_p are the Young's modulus and poisson's ratio of the plate material respectively. h is the thickness of the plate and a denotes the radius of the plate. It is found that when K_r > 100, the plate can be assumed to be rigid for all practical purposes. For the analyses herein, K_r = 10000 is used for the rigid plate and K_r = 1 is used for the flexible plate.

Fig. 8 shows the geometry of the plate-halfspace system with the boundary conditions and the location of the interface elements. A three noded axisymmetric triangular element is used to model the halfspace region while an

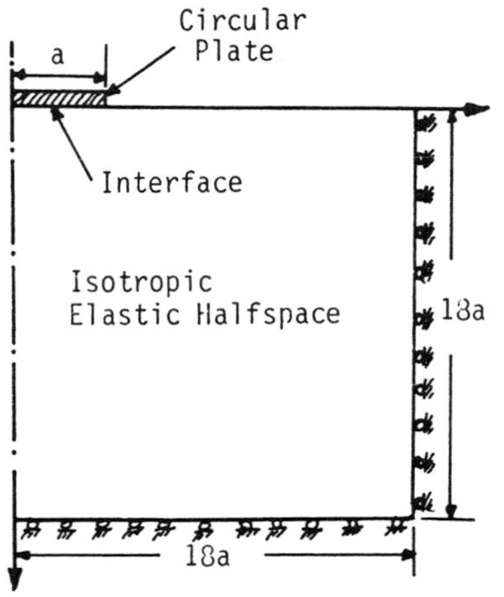

Fig. 8. Geometry of the plate-halfspace system with interfaces

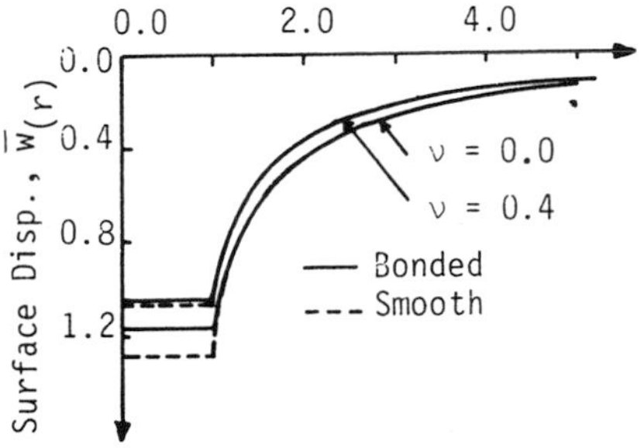

Fig. 9. Nondimensional displacement-radial distance response

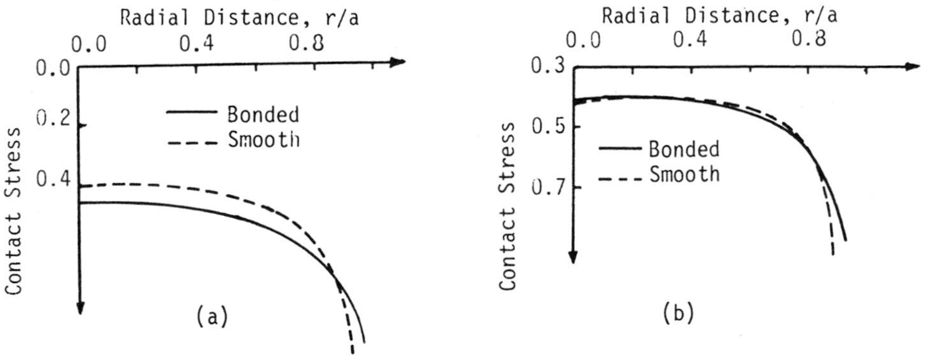

Fig. 10. Contact stress distribution beneath a rigid plate (a) for ν = 0.0. (b) for ν = 0.40

axisymmetric ring element is used for the plate (ref.6,11).

Fig. 9 shows the displacement of the surface of the halfspace as a function of the nondimensional radial distance for a rigid plate. Both perfectly bonded and perfectly smooth interface conditions are considered and results are presented for $\nu = 0.0$ and $\nu = 0.40$. Contact stress distributions beneath the plate are shown in Figs. 10a and 10b. It is seen that the effect of the interface

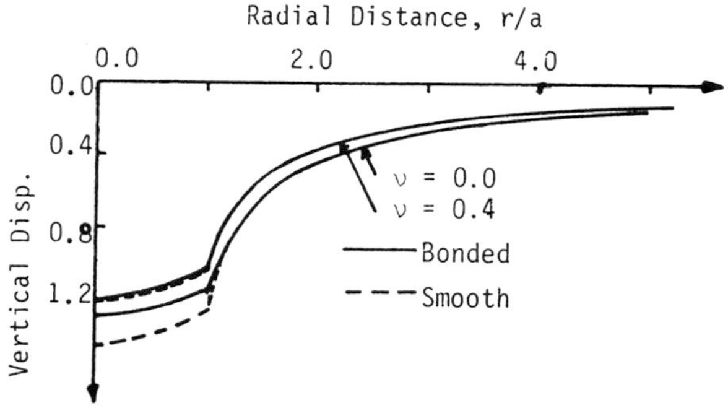

Fig. 11. Nondimensional displacement vs. radial distance for a flexible plate

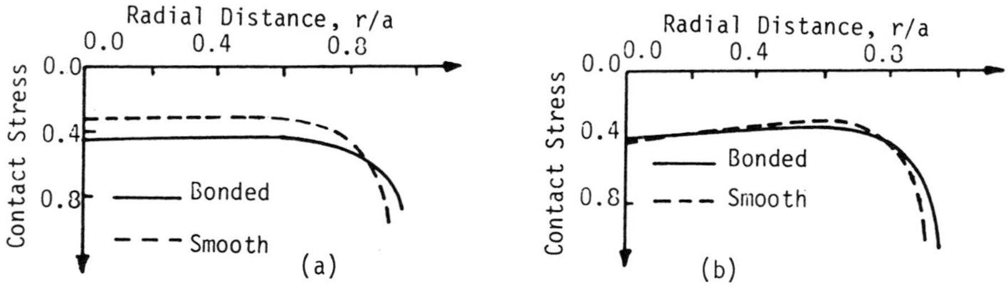

Fig. 12. Contact stress distribution beneath a flexible plate (a) for $\nu = 0.0$ (b) for $\nu = 0.40$

condition is maximum when $\nu = 0.0$. As $\nu \to 0.5$ (incompressible halfspace) the difference between these solutions tend to disappear.

Figs. 11, 12a and 12b show the corresponding results for a relatively flexible plate ($K_r = 1.0$). It is seen that for flexible plates the interface condition is more important than the rigid plates. This, however, tends to disappear as $\nu \to 0.5$.

Cylindrical inclusion within a halfspace

Anchors are often used in rock engineering and geotechnical engineering. An anchor is an inclusion problem and contains interface which may experience slip or opening during deformation.

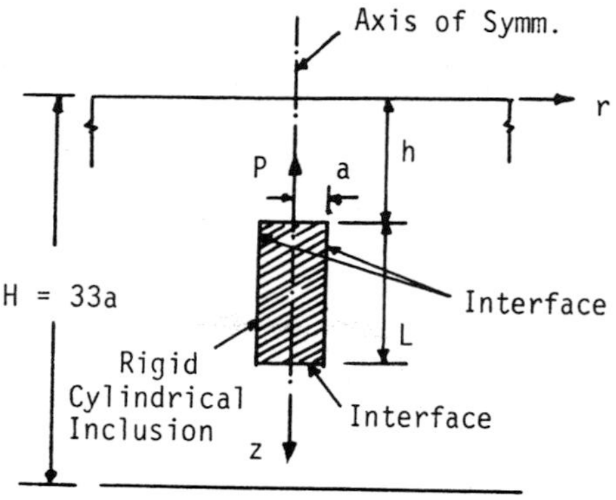

Fig. 13. Geometry of a cylindrical inclusion within an elastic halfspace

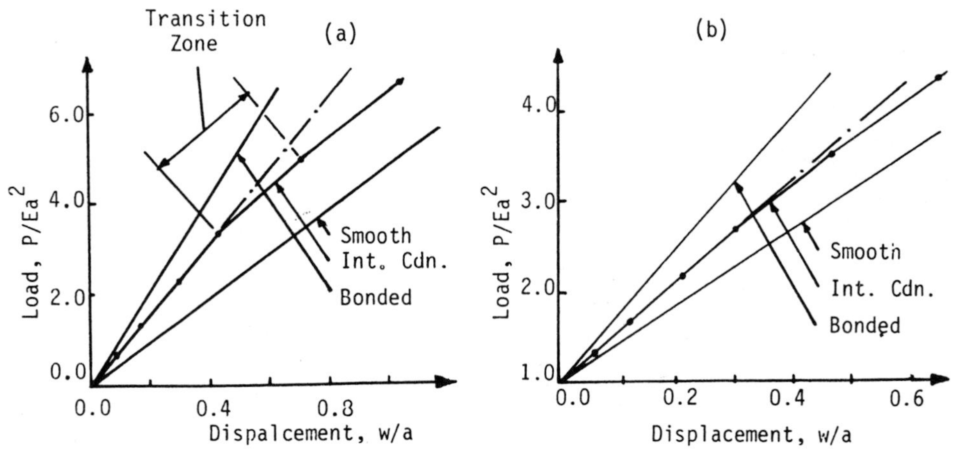

Fig. 14. Load vs displacement response curve for a rigid inclusion (a) for ν = 0.0 (b) for ν = 0.49

Fig. 13 shows the geometry and the location of the interface elements for a cylindrical inclusion within an elastic halfspace. The radius and the length of the inclusion are denoted by the symbols 'a' and 'L' respectively. The top face of the inclusion is at a distance of h from the halfspace surface. Fig. 13 also shows the applied load P at the top face of the cylinder.

The material constants used for the interface are $E_{nn} = 10^{10}$ psi and $\tau_y = 25$ psi. A parameter $\alpha_s = E_{ss}/E_{nn}$ is defined and results are presented for three different values of α_s i.e.

$\alpha_s = 0.0$ (perfectly smooth interface)
$\alpha_s = 1.0$ (perfectly bonded interface)
$\alpha_s = 10^{-10}$ (Intermediate interface condition)

Figs. 14a and 14b show the load-displacement response curves for the rigid inclusion. It is seen that the load displacement responses are significantly different for different interface conditions. For $\alpha_s = 10^{-10}$, however, a nonlinear load-displacement response is obtained. Again the effect of the interface condition is maximum when $\nu = 0.0$. Fig. 15 shows how the rigid displacement of the cylinder varies as a function of $-\ln(\alpha_s)$ for $\nu = 0.0$ and $\nu = 0.49$. Fig. 16 shows the distribution of the vertical stress on a horizontal section for z/a = 8.0 and for $\alpha_s = 0.0$ and $\alpha_s = 10^{-10}$. It is seen from Fig. 16 that the interface condition also changes the distributions of the stresses in the halfspace.

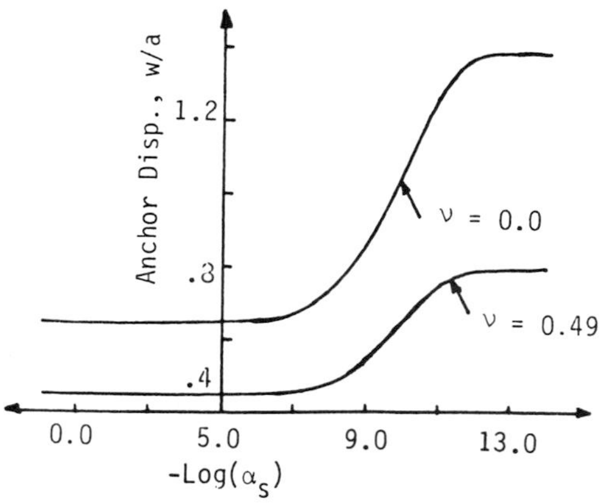

Fig. 15. Effect of α_s on the disp. for h/a = 6.0 and L/a = 5.0

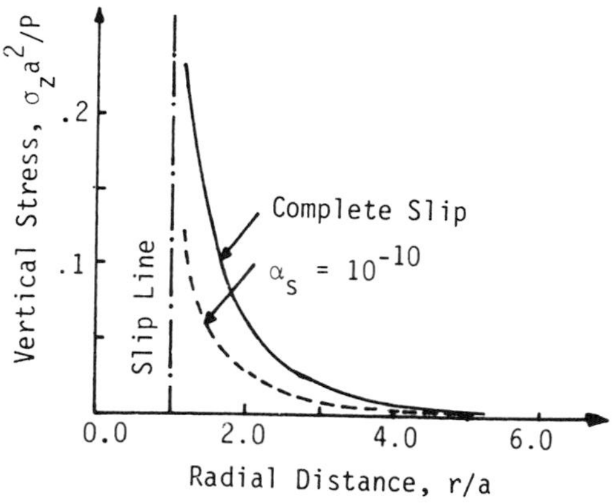

Fig. 16. Distribution of vertical stress for z/a = 8.0 and $\nu = 0.49$

CONCLUSIONS

A four noded axisymmetric interface element is used to simulate the slip and nonslip mode of shear deformation for material interfaces. The proposed interface element has finite thickness. Thus it can be easily used for modelling rock joints, faults, weakplanes, seams, etc. A parametric study is made to identify the important parameters which influences the rate of convergency of any iterative scheme. Two boundary value problems are analyzed and results are presented for various interface conditions. It is evident from the present analyses that the interface condition significantly affects the overall response of any boundary value problem. This is particularly important for soil-structure interaction problems.

REFERENCES

1 C.S. Desai and J.F. Abel, Introduction to the Finite Element Method, Van Nostrand Reinhold, New York, 1972.
2 C.S. Desai, Soil-Structure Interaction and Simulation Problems, in G. Gudehus (Ed.), Finite Elements in Geomechanics, 1977, 209 pp.
3 C.S. Desai, Behavior of Interfaces Between Structural and Geologic Media, A State-of-the-Art paper for Int. Conf. on Recent Advances in Geotech. Earthq. Eng. and Soil Dyn., St. Louis, Missouri, 1981.
4 C.S. Desai, M.M. Zaman, G.J. Lightner and H.J. Siriwardane, Thin-Layer Elements for Interfaces and Joints, Int. J. Num. Analyt. Meth. Geomech., Vol. 8, 1984, 19 pp.
5 M.O. Faruque and M.M. Zaman, Modelling of In Situ Tests in Jointed Rocks, Proc. Eighth Conf. on Electronic Computation, Houston, 1983, 88 pp.
6 M.O. Faruque, The Role of Interface Elements in Finite Element Analysis of Geotechnical Engineering Problems, M. Eng. Thesis, Carleton Univ., Canada, 1980.
7 R.E. Goodman, R.L. Taylor and T.L. Brekke, A Model for the Mechanics of Jointed Rocks, J. Soil Mech. Fdn. Div., ASCE, Vol. 94, 1968, 637 pp.
8 J. Ghaboussi, E.L. Wilson and J. Isenberg, Finite Element for Rock Joints and Interfaces, J. Soil Mech. Fdn. Div., ASCE, Vol. 99, 1973, 833 pp.
9 L.R. Herrmann, Finite Element Analysis of Contact Problems, J. Eng. Mech. Div., ASCE, Vol. 104, 1978, 1042 pp.
10 M.G. Katona, A Simple Contact-Friction Interface Element with Application to Buried Culvert, Proc. Impl. Comp. Procedures and Stress-Strain Laws in Geotech. Eng., C.S. Desai and S.K. Saxena (Eds.), Chicago, 1981, 45 pp.
11 J. Kirkhope and G.J. Wilson, Vibration and Stress Analysis of Thin Rotating Discs Using Annular Finite Elements, J. Sound and Vib., Vol. 44, 1976, 461pp.
12 M.A. Mahtab and R.E. Goodman, Three-Dimensional Finite Element Analysis of Jointed Rock Slopes, Proc. 2nd Cong. ISRM, Belgrade, Vol. 3, 1970.
13 A.P.S. Selvadurai and M.O. Faruque, The Influence of Interface Friction on the Performance of Cable Jacking Tests of Rock Masses, Proc. Impl. Comp. Pro. Stress-Strain Laws in Geotech. Eng., Chicago, Vol. 1, 1981, 169 pp.
14 A.P.S. Selvadurai, The Interaction Between a Uniformly Loaded Circular Plate and an Isotropic Elastic Halfspace: A Variational Approach, J. Struct. Mech., Vol. 7, 1979, 231 pp.
15 O.C. Zienkiewicz, B. Best, C. Dullage and K.G. Stagg, Analysis of Nonlinear Problems in Rock Mechanics with Particular Reference to Jointed Rock Systems, Proc. 2nd Cong. ISRM, Vol. 3, 1970, 501 pp.
16 M.M. Zaman, Ph.D. Dissertation, Univ. of Arizona, Tucson, 1982.

INTERFACE PHENOMENA IN FRACTURE MECHANICS

RAPID INTERFACE FLAW EXTENSION WITH FRICTION: VARYING COULOMB OR VISCOUS COEFFICIENTS OF FRICTION

L.M. Brock
Univ. Kentucky, Lexington, KY, 40506 (U.S.A.)

ABSTRACT

Interface flaw extension in compression and shear can lead to sliding crack surface contact with the possibility of friction. Two standard friction models - Coulomb and viscous - are studied for dynamic interface crack extension, and the results of analytical solutions indicate that the friction coefficients in both models should vary over the crack surface.

INTRODUCTION

Under appropriate loading, the bonding between adjacent layers in a laminated composite may fail, giving rise to a rapidly extending interface flaw, or crack, and stress wave propagation. In many situations varying from strata in the earth's crust to structural laminates, large ambient compressive stresses are present, so that the crack surfaces remain in sliding contact under friction.

Exact analyses (refs. 1,2) of a single growing crack at the interface of two half-planes under in-plane loading shows that a Coulomb friction law results in dynamic stress singularities which exceed the standard square-root value at one crack edge, thus implying an infinite energy release rate.

For viscous contact near the edges of a growing crack in an unbounded homogeneous material, analysis (ref. 3) shows that the stress singularities at both crack edges lie below the square-root value, and vanish in the limit as the crack edges approach the Rayleigh wave speed.

This analysis reconsiders the two friction model situations of refs. 1-3 for sliding contact over the whole interface crack. As before, the models are one-parameter (i.e. one friction coefficient is required). However, the coefficients are now not constant. The situations are formulated in the sense of Green's function problems (ref. 4). Moreover, the coefficient variation and crack speeds are of types which allow the problems to be dynamically similar. Nevertheless, the solution detail will be sufficient enough to allow some basic observations on the effect of the coefficient behavior on interface crack extension.

The Coulomb friction model problem is formulated in the next section and, subsequently, its solution by means of dislocation arrays is outlined. The viscous friction model is then considered. Finally, some basic aspects of the solution behavior which are independent of particular problem parameters and do not require numerical calculations are discussed.

FORMULATION OF COULOMB PROBLEM

Consider two isotropic, homogeneous, linearly elastic half-planes perfectly bonded along an interface defined in terms of the Cartesian coordinates (x,y) by y=0. When referring specifically to the half-plane y>0, the subscript 1 is affixed to the field variables and material parameters; the subscript 2 is employed when the half-plane y<0 is meant.

For $s<0$, where $s=v_M \times (\text{time})$ and v_M is the maximum wave speed in either half-plane, the half-planes are at rest. For $s>0$, a crack appears at the origin $(x,y)=0$ and grows along the interface in the positive (+) and negative (-) x-directions with the constant, subcritical speeds $v_{\pm}$. By subcritical, we mean that the speed lies below the minimum value of any wave speed in the half-planes. The crack surfaces remain in sliding contact which obeys a Coulomb friction law at every point. That is, for $y=0, -c_- s<x<c_+ s$

$$\sigma_{xy} - f(x/s)\sigma_y = \delta(x - c_o s), \quad (\sigma_1 - \sigma_2)\cdot \underset{\sim}{j} = (\underset{\sim}{u}_1 - \underset{\sim}{u}_2)\cdot \underset{\sim}{j} = 0 \qquad \text{(1a,b)}$$

where δ is the Dirac function, σ is the stress tensor with elements $(\sigma_x, \sigma_y, \sigma_{xy})$, $\underset{\sim}{j}$ is the y-basis vector, $(\cdot)$ denotes the inner product, $\underset{\sim}{u}$ is the displacement vector and

$$v_M c_{\pm} = v_{\pm}, \quad c_o < c_{\pm} \qquad (2)$$

No restrictions are at the outset placed on the Coulomb friction coefficient function f except that it be homogeneous of degree 0 in (x,s), and non-negative and bounded for all x/s. Along the interface portion exterior to the crack, perfect bonding requires that

$$(\sigma_1 - \sigma_2)\cdot \underset{\sim}{j} = \underset{\sim}{u}_1 - \underset{\sim}{u}_2 = 0 \qquad (3)$$

In each half-plane, it is convenient to follow the precedent of (2) and to introduce the dimensionless speed-related quantities

$$n = v_M/v_d, \quad m = v_d/v_r \qquad (4)$$

where (v_d, v_r) are the dilatational and rotational wave speeds, respectively. It is then easily shown that the governing equations in each half-plane take the form

$$\nabla^2 \underset{\sim}{u}+(m^2-1)\underset{\sim}{\nabla}(\underset{\sim}{\nabla}\cdot\underset{\sim}{u})-m^2n^2\ddot{\underset{\sim}{u}} = 0 \tag{5}$$

$$\frac{1}{\mu}\sigma = (m^2-2)II+\underset{\sim\sim}{\nabla u}+\underset{\sim\sim}{u\nabla} \tag{6}$$

where $\underset{\sim}{\nabla}$ and ∇^2 are the gradient and Laplacian operators in the (x,y)-plane, $(\dot{\ })\equiv\partial(\ )/\partial s$, μ is the shear modulus and II is the identity tensor. Finally, the initial conditions on the problem are that for $s\leq 0$

$$\underset{\sim}{u}\equiv 0. \tag{7}$$

DISLOCATION ARRAYS

It is noted that the solution to the previous problem will be dynamically similar because $\underset{\sim}{u}$ will clearly be homogeneous of degree 0 in (x,y,s). This, in turn, follows from the lack of a characteristic length in the problem owing to the constant crack speeds and homogeneity of f. This observation suggests that we first consider the related problem of an array of dislocations extending both by climb and glide along the region of the interface corresponding to the crack. That is, eq. (1) is replaced by

$$(\sigma_1-\sigma_2)\cdot\underset{\sim}{j} = 0,\ \underset{\sim}{u}_1-\underset{\sim}{u}_2 = \int_{-k_-}^{t} \underset{\sim}{B}(k)dk \tag{8}$$

where

$$t = s/x,\ k_\pm = 1/c_\pm \tag{9}$$

and $\underset{\sim}{B}=(B_x,B_y)$ is a dislocation array function which satisfies the requirement

$$\int \underset{\sim}{B}(k)dk = 0 \tag{10}$$

Here $\int(\ )dk$ denotes integration over the range $(-k_-,k_+)$ so that eq. (10) implies that $\underset{\sim}{u}_1=\underset{\sim}{u}_2$ at the dislocation edges. Following the work of ref. 5, the problem defined by eqs. (2)-(10) can be solved by elementary Fourier-Laplace transform methods (ref. 6) coupled with Cagniard-deHoop (ref. 7) transform inversions. It is then easily shown that the two interface traction elements in the dislocation array region have the forms

$$|x|\sigma_{xy} = -\frac{A(t)}{\pi t} ⨍ \frac{B_x(k)}{t-k}\,kdk - C(t)B_y(t) \tag{11}$$

$$|x|\sigma_y = -\frac{B(t)}{\pi t} ⨍ \frac{B_y(k)}{t-k} kdk + C(t)B_x(t) \tag{12}$$

for s>0, where ⨍ denotes Cauchy Principal Value integration and

$$S(t)A(t) = h(R_1 m_2^2 n_2^2 a_2 + R_2 m_1^2 n_1^2 a_1),\ h\mu_1 = \mu_2 \tag{13}$$

$$S(t)B(t) = h(R_1 m_2^2 n_2^2 b_2 + R_2 m_1^2 n_1^2 b_1) \tag{14}$$

$$S(t)C(t) = ht(R_1 L_2 - R_2 L_1) \tag{15}$$

$$S(t) = 4(1-h)^2 t^2 a_1 b_1 a_2 b_2 - hm_1^2 n_1^2 m_2^2 n_2^2 (a_1 b_2 + a_2 b_1)$$

$$- (2ht^2 - T_1)^2 a_2 b_2 - (2t^2 - hT_2)^2 a_1 b_1 + t^2 (T_1 - hT_2)^2 \tag{16}$$

$$T = m^2 n^2 - 2t^2,\ R = \mu(4t^2 ab + T^2),\ L = 2ab + T \tag{17}$$

$$a = \sqrt{(t^2 - n^2)},\ b = \sqrt{(t^2 - m^2 n^2)} \tag{18}$$

The functions R are the Rayleigh functions, with zeros at $t=\pm m_R$, where $v_R = v_M/m_R$ is the Rayleigh wave speed, and $m_R > m$. The function S is the Stoneley wave function, which has zeroes only when (ref. 8)

$$S(m_2 n_2) > 0\ (m_2 n_2 > m_1 n_1),\ S(m_1 n_1) > 0\ (m_1 n_1 > m_2 n_2) \tag{19}$$

These zeroes occur at $t=\pm m_S$, where $v_S = v_M/m_S$ is the Stoneley wave speed, and $m_S > (m_1, m_2)$.

REDUCTION OF COULOMB PROBLEM

Clearly, the dislocation array problem will also exhibit $\tilde{u}$-fields which are homogeneous of degree 0 and, indeed, will give the solution to the Coulomb problem provided that $B_y \equiv 0$ and a B_x can be found that allows eq. (1) to be satisfied. Upon introducing the convenient variable changes

$$k^2 \tilde{B}(k) = \tilde{b}(1/k),\ t^2 A(t) = \alpha(1/t),\ t^2 C(t) = \beta(1/t) \tag{20}$$

in eqs. (8)-(18), therefore, the Coulomb problem can be reduced to the equation

$$\frac{\alpha(z)}{\pi} \fint \frac{b_x(c)}{c-z} dc + \beta(z)f(z)b_x(z) = -|x|\delta(x-c_0 s) \tag{21}$$

for $-c_- < z < c_+$ under the constraint

$$\int b_x(c)dc = 0 \tag{22}$$

where $\int(\;)dc$ denotes integration over the range $(-c_-, c_+)$ and

$$z = x/s = 1/t \tag{23}$$

Equation (21) is a Fredholm integral equation of the first kind with a Cauchy kernel. It's solution is readily found to be (ref. 9)

$$b_x(z) = -\frac{K(z)}{L^2(z)}|x|\alpha(z)\delta(x-c_0 s) + M(z) \tag{24}$$

$$M(z) = F(z)\left[\frac{b_0\alpha(z)}{c_-+z} + \frac{1}{\pi}\frac{1}{c_0-z}\frac{1}{F(c_0)L^2(c_0)}\right] \tag{25}$$

where the constant b_0 follows from eqs. (22) and (24) as the solution to

$$\int M(c)dc = \frac{K(c_0)\alpha(c_0)}{L^2(c_0)} \tag{26}$$

In eqs. (24)-(26)

$$L(z) = \surd[1+K^2(z)], \quad K(z) = \frac{\beta(z)f(z)}{\alpha(z)} \tag{27}$$

$$F(z) = \left(\frac{c_+-z}{c_-+z}\right)^{-\frac{1}{2}+q(z)} \frac{\alpha(z)}{L(z)} e^{Q(z)} \tag{28}$$

$$\pi q(z) = \tan^{-1}K(z), \quad Q(z) = \int \frac{dc}{c-z}[q(c)-q(z)] \tag{29}$$

VISCOUS FRICTION PROBLEM SOLUTION

Consider the same situation as for the Coulomb friction problem, except that now eq. (1a) is modified to read

$$\sigma_{xy} - f(x/s)\underset{\sim}{i}\cdot(\dot{\underset{\sim}{u}}_1 - \dot{\underset{\sim}{u}}_2) = \delta(x-c_0 s) \tag{30}$$

where $\underset{\sim}{i}$ is the x-basis vector and the viscous friction coefficient $f(x/s)$ is constrained at the outset to be homogeneous of degree 0 in (x,s), non-negative, and bounded for all x/s. It is readily shown that the solution to this problem can also be reduced by dislocation arrays to the equation set (21) and (22), except that the definition of β is now

$$\beta(z) = z^2 f(z)\,\mathrm{sgn}(z) \tag{31}$$

Therefore, the solution to this problem has the basic form given by eqs. (24) - (26).

EFFECTS OF COEFFICIENT BEHAVIOR

Some important aspects of the interface crack extension process can be ascertained in the behavior of the stress fields near the crack edges. We here focus attention on the shear tractions, since they govern the crack formation process. It is readily shown for both friction models that

$$|x|\sigma_{xy} \sim O(|z-c_+|^{-\frac{1}{2}+q(c_+)}),\ |x|\sigma_{xy} \sim O(|z+c_-|^{-\frac{1}{2}+q(-c_-)}) \tag{32}$$

for $x \to c_+ s^+$ and $x \to -c_- s^-$, respectively. That is, the order of the crack edge singularity is a function of crack edge speed, a result also noted in refs. 1-3.

A study of eq. (28) in view of eq. (21) shows that, for the Coulomb model in general,

$$q(c_+)/q(-c_-)<0,\ |q(c)|<1/2 \tag{33}$$

when $|c|, c_\pm$ are subcritical and f is constant. This result implies in view of eq. (32) that one crack edge stress field is less than square-root singular, while the other is integrable but greater than square-root singular. The choice of material parameters (m,n,μ,h) determines which crack edge is which (ref. 2). The crack edge with the larger singularity will also have an unbounded energy release rate associated with it, since this rate is a quadratic function of stress. This unfortunate result would seem to cast doubt on the validity of a Coulomb friction model for sliding contact of interface crack surfaces.

The result is, as noted above, somewhat general for, as the results of ref. 2 show, there do exist special sets of (m,n,μ,h) for which subcritical values of c can be found such that

$$q(c) = 0 \tag{34}$$

That is, certain half-plane combinations give square-root singular crack edge stress fields and, thus, physically reasonable energy release rates.

A more satisfactory result, however, can be obtained simply by allowing, as in the present analysis, the parameter f to vary. Then, one can guarantee eq. (34) at the crack edges by requiring that

$$f(\pm c_{\pm}) = 0 \tag{35}$$

This idea can also be viewed as follows:

(1) The Coulomb model, strictly speaking, requires proportionality between the resultant normal and tangential contact forces over a contact region; the requirement of point-wise proportionality may be severe.

(2) As crack surface is generated and subsequently subjected to time-varying traction fields, its frictional properties (i.e. f) may change.

(3) At the point of fracture, it may be more reasonable within the context of a continuum model, to expect the stress field on the crack surface side to continuously achieve a finite value (i.e. the crack surfaces to come into sliding contact).

(4) A continuously varying f may partially free the friction model from the necessity of detailed and perhaps experimentally unverifiable schemes of contact zones with differing constant friction coefficients.

A study of eq. (29) in view of eq. (31) shows that, for the viscous friction model,

$$0<q(c)<1/2 \tag{36}$$

when $|c|$ is subcritical, no Stoneley waves exist and f is constant. If, however, Stoneley waves do exist, that is, eq. (19) is satisfied, then we find

$$-1/2<q(c)<0 \tag{37}$$

Equations (36) and (37) imply that, if no Stoneley waves exist, then both crack edge stress fields are less than square-root singular while, if they do exist, then both fields are greater than square-root singular, with the aforementioned attendant difficulties.

Again, therefore, the validity of an apparently reasonable model depends on the choice of material parameters. A satisfactory result can, however, be obtained in general by allowing the coefficient f to vary in such a way that eq. (35) is satisfied. Items (2)-(4) discussed above can also be marshalled in favor of such a model for viscous friction.

STRESS INTENSITY FACTORS

If the models suggested above are adopted, then it is easily shown that the shear stresses near the crack edges give dynamic stress singular behavior of the form

$$|x|\sigma_{xy} \sim \frac{K_0}{\sqrt{(z-c_+)}}, \quad K_0 = \frac{b_0\alpha(c_+)e^{Q(c_+)}}{\sqrt{(c_+ + c_-)}} \tag{38}$$

$$|x|\sigma_{xy} \sim \frac{K_0}{\sqrt{|z+c_-|}}, \quad K_0 = \frac{b_0\alpha(-c_1)e^{Q(-c_-)}}{\sqrt{(c_+ + c_-)}} \tag{39}$$

for $x \to c_+ s^+$ and $x \to -c_- s^-$, respectively, where of course

$$Q(\pm c_\pm) = -\int \frac{q(c)}{c-z}\, dc \tag{40}$$

It is noted that the singularity arises from the b_0-term in eq. (24), which is actually the solution to the homogeneous version of eq. (21) (ref. 9)

CONCLUSIONS

The results presented here indicate that a variable coefficient of Coulomb or viscous friction which vanishes continuously at the edges of an interface crack allows two standard models of sliding contact to avoid the possibility of an undefined energy release rate. Moreover, a variable coefficient in its own right has validity in terms of a continuum explanation of fracture at the crack.

The results are presented in terms of Green's function types of problems, and the problems are dynamically similar because the coefficients are homogeneous functions of degree 0, and the crack speeds are constant. Moreover, certain auxiliary conditions (refs. 1,2,10) of sliding contact have not been examined; in particular, the necessity of non-tensile normal tractions on the interface crack surfaces in both models, and the consistency between the shear traction and relative slip directions in the Coulomb model. Nevertheless, the results presented here provide some potential for further development and conjecture.

ACKNOWLEDGEMENT

This research was partially supported by NSF Grant ENG7811773.

REFERENCES

1 L.M. Brock, in Advances in Engineering Science, Vol. 1, NASA CP-2001(1978) pp. 239-246
2 L.M. Brock, Int. J. Engng. Sci. 17(1979) pp. 49-58
3 L.M. Brock, Int. J. Engng. Sci. 20(1982) pp. 663-672
4 I. Stakgold, Green's Functions and Boundary Value Problems, Wiley-Interscience, New York, 1979, 638 pp.
5 L.M. Brock and Y.C. Deng, Int. J. Engng. Sci. 23(1985) pp. 163-171
6 I.N. Sneddon, The Use of Integral Transforms, McGraw-Hill, New York, 1972, 539 pp.
7 A.T. deHoop, Appl. Sci. Res. B8(1960) p 49-60
8 L.M. Brock and J.D. Achenbach, in Developments in Theoretical and Applied Mechanics, Vol. 5 (1971)
9 G.F. Carrier, M. Krook and C.E. Pearson, Functions of a Complex Variable, McGraw-Hill, New York, 1966, 438 pp.
10 L.M. Brock, in Fracture Mechanics in Ceramics, Vol. 5, Plenum, New York, 1983, 692 pp.

STUDY OF FRICTIONAL CLOSURE OF A CRACK EMANATING FROM A CIRCULAR INCLUSION

M. COMNINOU and N. LI

Dept. of Mechanical Engineering and Applied Mechanics, University of Michigan, Ann Arbor MI 48109-2125 (USA)

ABSTRACT

Consider a radial crack emanating from an elastic inclusion or fiber in a matrix under uniaxial tension. We study the various parameters that affect crack closure and allow for frictional forces to be transmitted by the closed crack faces. Representative examples are given for special cases.

INTRODUCTION

Interest in the fracture of fiber-reinforced composites has motivated various studies dealing with the interaction of cracks and inclusions. Erdogan and his co-workers have considered several configurations involving cracks in the presence of inclusions in the early seventies. The references pertinent to the present study are by Atkinson (ref.1) and by Erdogan and Gupta (ref.2), because they also consider radial cracks. Unlike all previous work on the subject which is valid only when the crack remains completely open, we are especially considering the possibility of partial or complete closure and seek to establish the material combinations and geometry for which it occurs. In addition we allow frictional forces to be transmitted through the closed crack faces and consequently have to study regions of stick, slip and separation in general. The special case of a hole has already been considered (ref.3).

Although a complete parametric study is currently under way, only a few representative results are given here.

FORMULATION

The formulation of the problem is along the same lines as ref.3. We consider the general case for which there are three zones along the crack: a zone of stick , R<r<a, a zone of slip, a<r<b, and a zone of separation, b<r<c. The crack is inclined by an angle θ with respect to the applied tension T, as shown in Fig. 1.

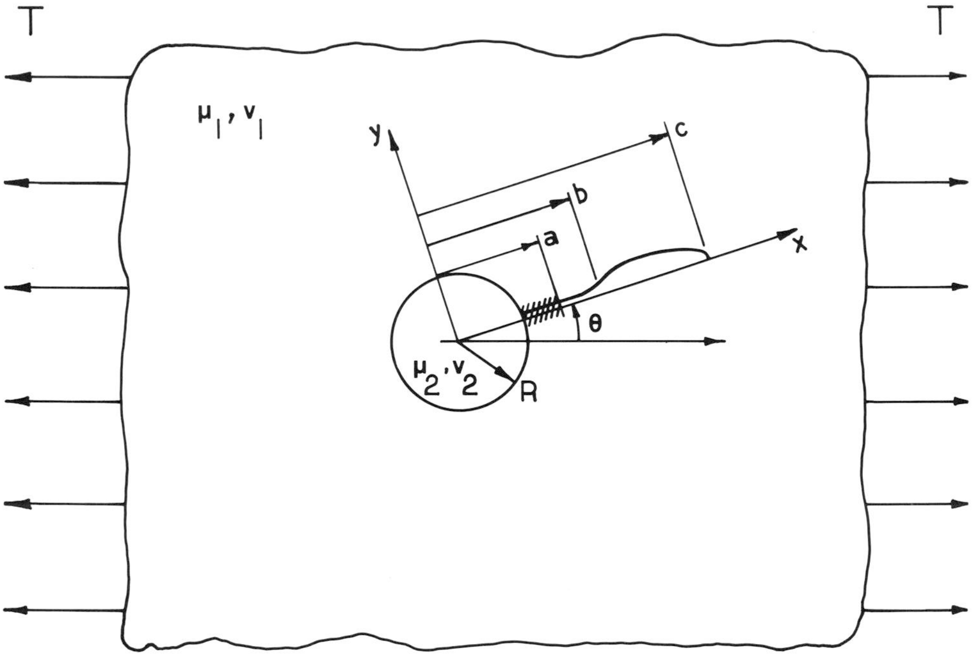

Fig. 1. Geometry of the problem.

First we consider the tractions along radial lines in the absence of the crack (ref.4)

$$\sigma_{yy}(x) = \frac{T}{2}\left\{1 - \frac{2\beta}{1+\alpha-2\beta}\frac{R^2}{r^2} - \left[1 - \frac{3(\alpha-\beta)}{1+\beta}\frac{R^4}{r^4}\right]\cos 2\theta\right\} \quad (1)$$

$$\sigma_{xy}(x) = \frac{T}{2}\left[-1 + \frac{2(\alpha-\beta)}{1+\beta}\frac{R^2}{r^2} - \frac{3(\alpha-\beta)}{1+\beta}\frac{R^4}{r^4}\right]\sin 2\theta \qquad (2)$$

where

$$\alpha = \frac{\Gamma(\kappa_1+1) - (\kappa_2+1)}{\Gamma(\kappa_1+1) + (\kappa_2+1)}, \quad \beta = \frac{\Gamma(\kappa_1-1) - (\kappa_2-1)}{\Gamma(\kappa_1+1) + (\kappa_2+1)} \qquad (3)$$

$$\Gamma = \mu_2/\mu_1 \;; \quad \kappa = 3 - 4\nu \text{ for plane strain} \qquad (4)$$

and μ and ν denote shear modulus and Poisson's ratio respectively. Subscripts are used to denote material 1 (matrix) or material 2 (fiber).

To account for the partially closed frictional crack we introduce a climb dislocation distribution B_y over the separation zone and a glide dislocation distribution B_x over the slip zone. The resulting tractions on $y = 0$ can be obtained from Dundurs and Mura (ref.5) and are

$$\tau_{yy}(x) = \frac{2\mu_1}{\pi(\kappa_1+1)}\int_b^c B_y(\xi)\, K_y(x,\xi)\,d\xi \qquad (5)$$

$$\tau_{xy}(x) = \frac{2\mu_1}{\pi(\kappa_1+1)}\int_a^c B_x(\xi)\, K_x(x,\xi)\,d\xi \qquad (6)$$

where

$$\begin{aligned} K_y(x,\xi) = {} & \frac{1}{x-\xi} + \frac{\alpha+\beta^2}{1-\beta^2}\frac{\xi}{x\xi - R^2} \\ & + \frac{\alpha-\beta}{1+\beta}\frac{(\xi^2-R^2)R^2}{\xi(x\xi-R^2)^2}\left[\frac{\xi^2}{R^2} - \frac{\xi^2-R^2}{x\xi-R^2}\right] \\ & + \frac{R^2}{2\xi x^2}\left[\frac{\beta-\alpha}{1+\beta}\frac{2\xi^2-R^2}{R^2} + \frac{1-\alpha^2}{(1-\beta)(1+\alpha+2\beta)} - 1\right] \\ & - \frac{\alpha+\beta^2}{1-\beta^2}\frac{1}{x} - \frac{\alpha-\beta}{1+\beta}\frac{R^2}{x^3} \end{aligned} \qquad (7)$$

$$K_x(x,\xi) = \frac{1}{x-\xi} + \frac{\alpha+\beta^2}{1-\beta^2}\frac{\xi}{x\xi - R^2}$$

$$+ \frac{\alpha-\beta}{1+\beta}\frac{(\xi^2-R^2)R^2}{\xi(x\xi-R^2)^2}\left[1 - \frac{\xi^2-R^2}{x\xi-R^2}\right]$$

$$- \frac{(1+\alpha)\beta}{1-\beta^2}\frac{R^2}{\xi x^2} - \frac{\alpha+\beta^2}{1-\beta^2}\frac{1}{x} + \frac{\alpha-\beta}{1+\beta}\frac{R^2}{x^3} \qquad (8)$$

The dislocation distributions must satisfy the following conditions of uniqueness of displacements

$$\int_b^c B_y(\xi)d\xi = 0 \qquad \int_a^c B_x(\xi)d\xi = 0 \qquad (9,10)$$

Denoting the total normal and shear tractions by $N(x)$, $S(x)$ on $y = 0$ we have

$$N(x) = \sigma_{yy}(x) + \tau_{yy}(x) \qquad (11)$$

$$S(x) = \sigma_{xy}(x) + \tau_{xy}(x) \qquad (12)$$

The boundary conditions are now

$$N(x) = 0 \quad b < x < c \qquad (13)$$

$$N(x) \leq 0 \quad a < x < b \qquad (14)$$

$$S(x) = -f\,\mathrm{sgn}\,h(x)N(x) \quad , \quad a < x < c \qquad (15)$$

$$|S(x)| < f\,|N(x)| \quad , \quad R < x < a \qquad (16)$$

where $h(x)$ is the tangential shift and

$$B_x(x) = -\frac{dh(x)}{dx} \qquad (17)$$

and f is the coefficient of friction between the crack faces. Note that the extent of gap and the climb dislocation distribution can be determined independently of friction and shear considerations. The results are then used to determine slip and shear tractions. Each step involves the numerical solution of a Cauchy type singular integral equation with an auxiliary condition. The method of Erdogan et al was used for the discretization of the singular equations (ref. 6). The inequalities of the problem are verified a posteriori. For more details in the computational aspects the interested reader is referred to ref. 7. When slip penetrates to the left tip a = R, the asymptotic analysis of ref. 2 must be used.

RESULTS

It is convenient to present most of the results in the plane of the composite parameters α, β, ref. 8. Fig. 2 shows the loci of the completely open crack boundary for various angles   and fixed c/R = 4. The completely open region is below each curve. Fig. 3 shows the loci of the completely closed crack boundary. The crack is completely closed above each curve for c/R = 4. The curves stop when a gap detached from the crack tips is initiated. Fig. 4 shows the loci of b = const for c/R = 4 and θ = 10°. The curves stop at points where a second gap becomes possible. In the same figure the boundary curve for which the slip zone reaches the crack tip, a = R, is also shown for f = .5. Figures 5, 6, and 7 present similar results but for a shorter crack, c/R = 2. An

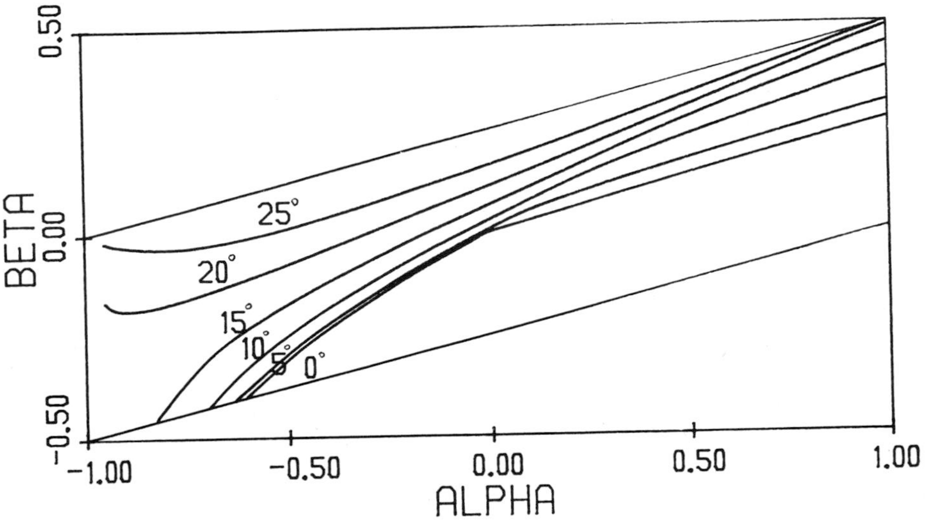

Fig. 2. Loci of the completely open crack, c/R = 4.

example of the gap shape and the normal tractions in the contact zone is shown in Fig. 8 . In this case slip is about to penetrate to the left tip, a = R.

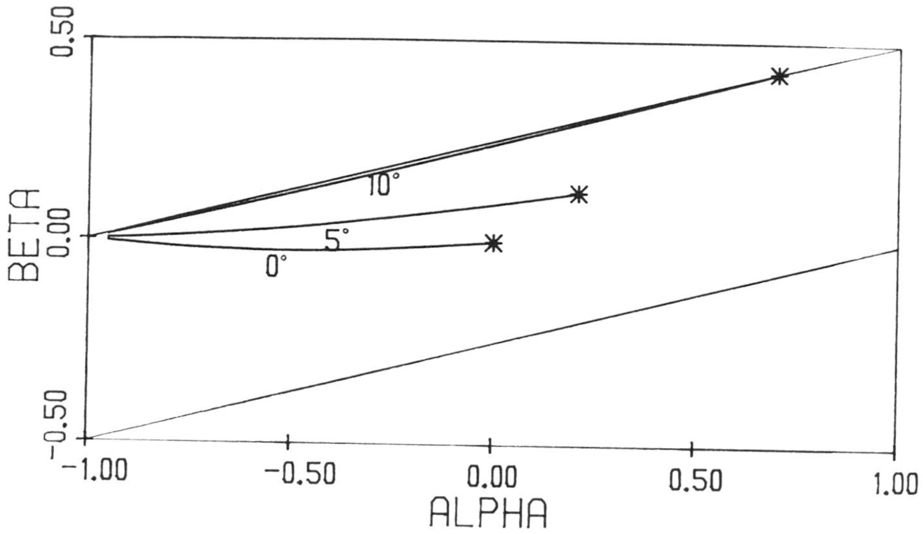

Fig. 3. Loci of the completely closed crack, c/R = 4.

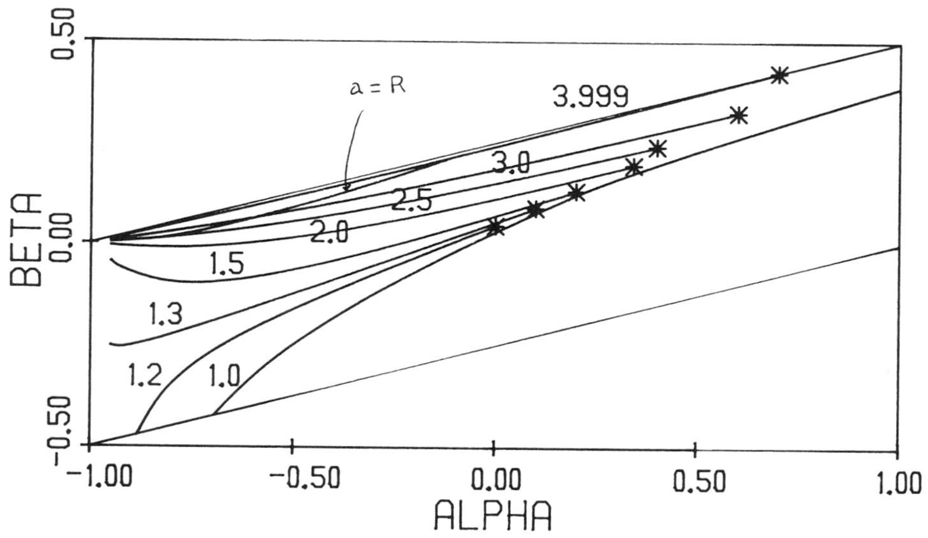

Fig. 4. Loci of constant b for c/R =4 and θ = 10°.

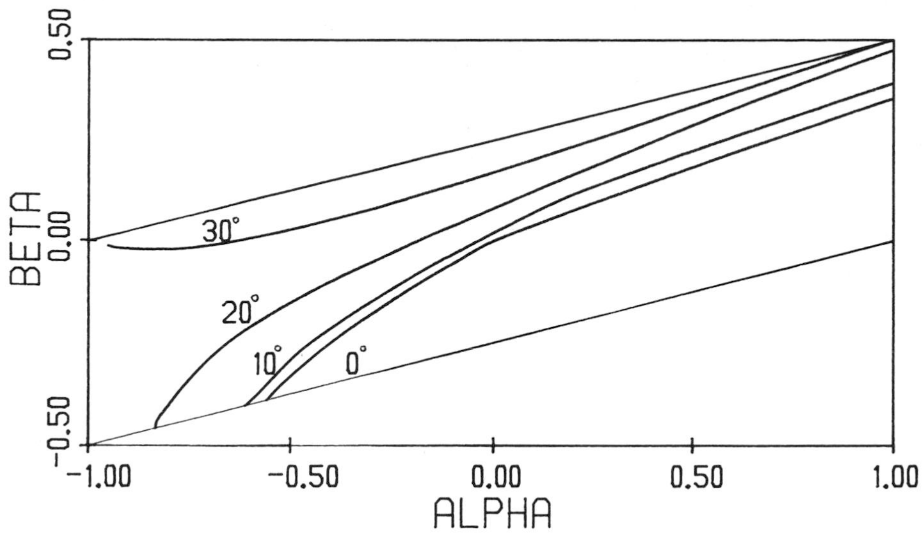

Fig. 5. Loci of the completely open crack, c/R = 2.

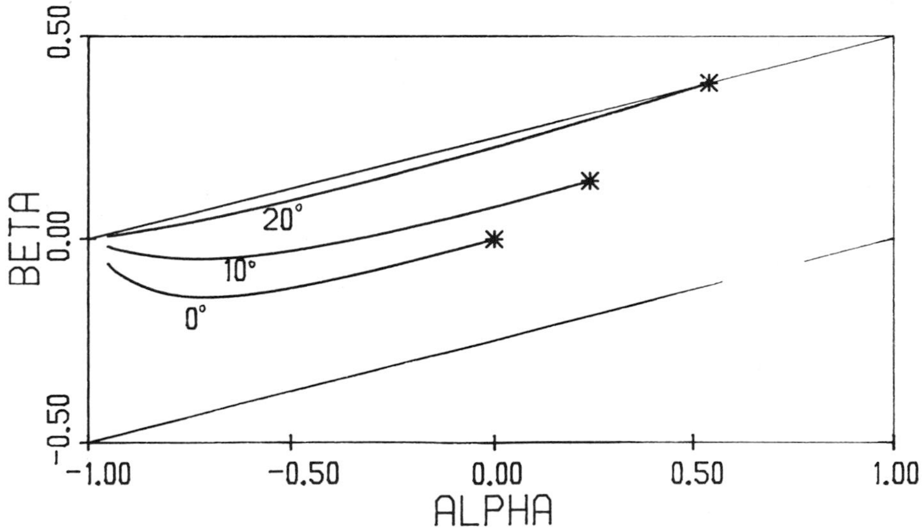

Fig. 6. Loci of the completely closed crack, c/R = 2.

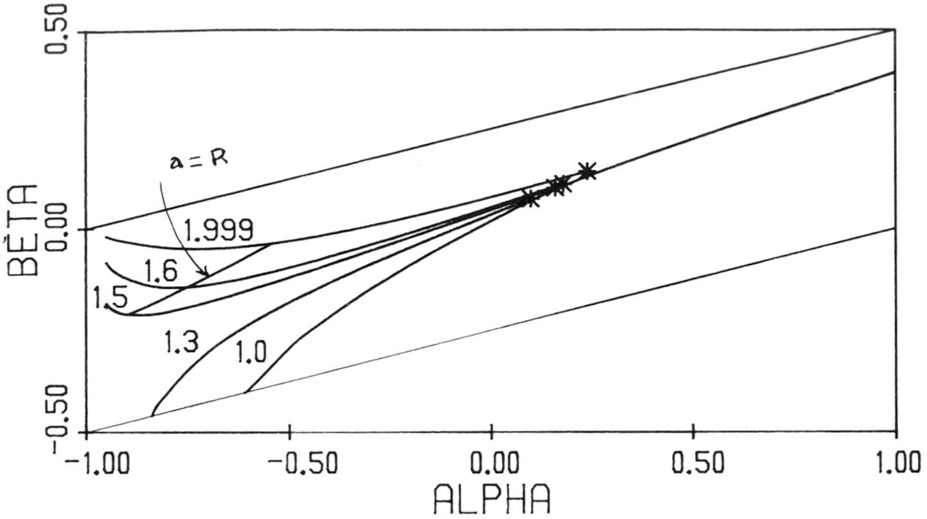

Fig. 7. Loci of constant b for c/R = 2 and $\theta = 10^{\circ}$.

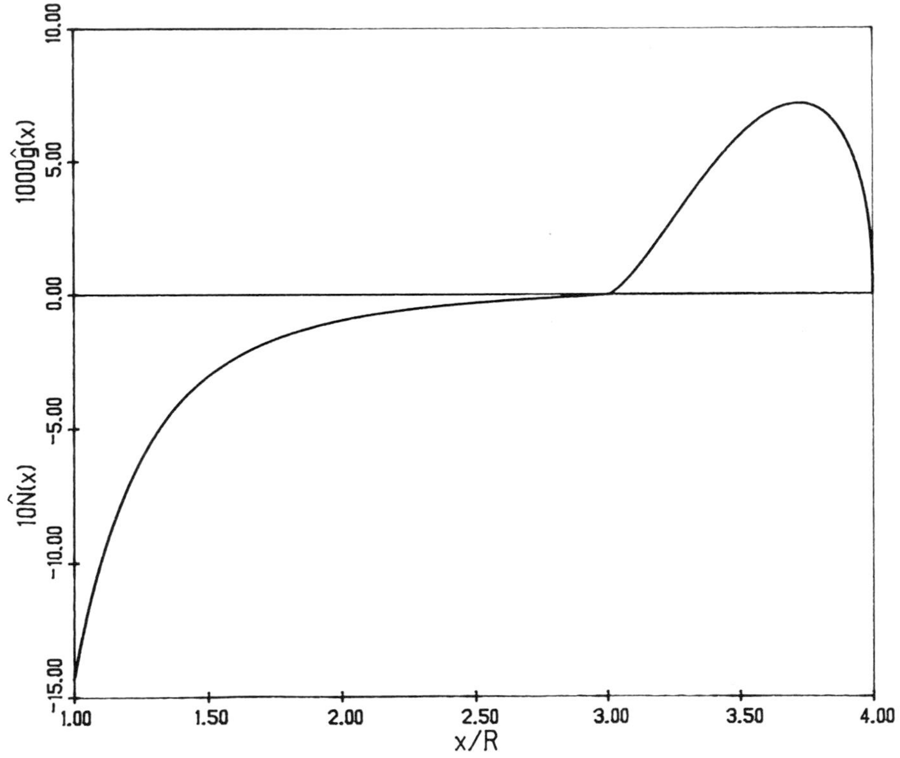

Fig. 8. Gap profile and normal tractions for c/R = 4, b/R = 3, f = .5, and $\theta = 10^{\circ}$.
$\hat{N}(x) = \hat{N}(x)/T$, $\hat{g}(x) = 4\mu_1 \hat{g}(x)/(\kappa_1+1)TR$.

CONCLUSION

It has been shown that, depending on the inclination of the radial crack with respect to the applied uniform tension, the crack faces may be partially closed and under conditions of slip. It is possible to have a partial gap next to the right crack tip, a gap detached from the crack tips or both. Depending on the condition on which a crack tip is we may or may not have a stress intensity factor in mode I or II. A comprehensive parametric study of the stress intensity factors will appear elsewhere.

ACKNOWLEDGEMENT

Partial support by the Office of Naval Research through contract N00014-84-K-0287 is gratefully acknowledged.

REFERENCES

1 C. Atkinson, Int. J. Engng. Sci., 10 (1972) 127-136.
2 F. Erdogan and G.D. Gupta, Int. J. Fract., 11 (1975) 11-27.
3 M. Comninou and F-K Chang, Int. J. Fract., in press.
4 J.N. Goodier, Trans. ASME, 55 (1933) 39-44.
5 J. Dundurs and T. Mura, J. Mech. Phys. Solids, 12 (1964) 177-189.
6 F. Erdogan, G.D. Gupta and T.S. Cook, in G.C. Sih (Ed.), Methods of Analysis and Solution of Crack Problems, Noordhoff, 1973, pp. 369-423.
7 D.A. Hills and M. Comninou, Int. J. Solids Struct., 21 (1985) 399-410.
8 J. Dundurs, in T. Mura (Ed.), Mathematical Theory of Dislocations, ASME, 1969, pp. 70-115.

FINITE ELEMENT INTERFACE CRACK ANALYSIS WITH APPLICATION TO MASSIVE HYDRAULIC FRACTURING CONTAINMENT*

E. P. Chen
Applied Mechanics Division III, Sandia National Laboratories, P.O. 5800, Albuquerque, NM 87185 (USA)

ABSTRACT

The hydraulic fracturing containment problem is idealized to the analysis of an interface crack problem. Unlike previous studies, the interface crack geometry with the correct crack tip singularity was modeled. To accomplish this purpose, analytical interfacial crack tip stress and displacement expressions were derived and implemented into the finite element code PAPST via the enriched element technique. The accuracy of this technique was validated through comparisons with analytical and other numerical solutions in previous studies. Parametric studies were then carried out to determine the relative importance of in situ stress and material property difference on fracture containment. A Maximum Stress Fracture Criterion was defined for this purpose The results indicated that in situ stress and material property difference play different roles in containing crack growth through the material interface. Prior to the arrival of the crack tip at the interface, material property difference dominates the containment process by determining the pressure level required for the crack to grow to the interface. Once the crack tip arrives at the interface, the in situ stress is the dominant factor in containing crack growth through the material interface.

INTRODUCTION

A significant question associated with the application of the massive hydraulic fracturing process for unconventional gas recovery purposes involves the containment of the fracture propagation in the pay zone. This problem can be described in terms of the schematic diagram as shown in Figure 1. Optimum pressurization from the borehole such that uniform pressure condition is achieved on the crack surface is assumed. Hence, the crack front extends in a circular manner in the pay zone characterized by material 1 (typically Sandstones). If sufficient pressure is maintained in the borehole, the vertical growth will eventually approach the material interface between material 1 and 2 (typically shales). If this vertical growth cannot be contained in material 1, the massive hydraulic fracturing process becomes cost-

* This work was sponsored by the Department of Energy's Western Gas Sand Subprogram, currently managed by Morgantown Energy Technology Center.

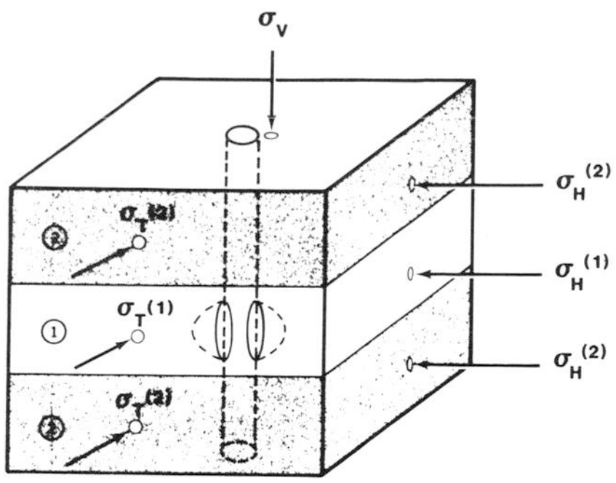

Figure 1. Schematic Diagram of the Hydraulic Fracturing Process

ineffective since a large amount of the pressure producing volume will be wasted in the fracturing of the nonpaying material. The problem involves three-dimensional geometries and is very complex in nature. A tractable approach to the containment question can be made by treating simplified plane strain geometries, Figure 2. This model has been used by many researchers [e.g. 1–6] to discuss the problem. The primary parameters have been identified as the in situ stresses (Figures 1 and 2) and the mechanical properties of the materials associated with the interface. The interaction between these

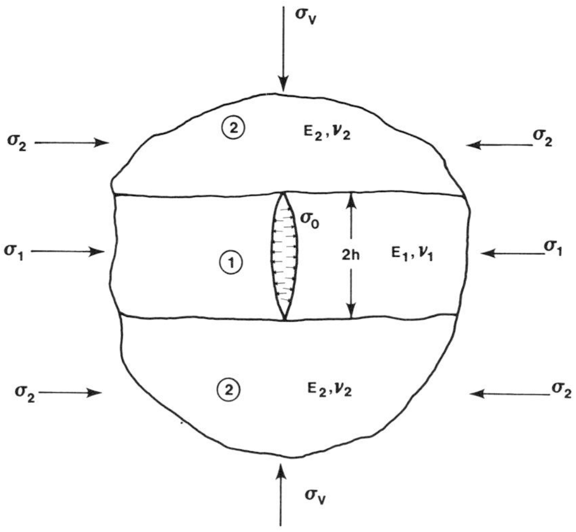

Figure 2. Simplified Plane Strain Interface Crack Geometry

parameters and the interface crack geometry determines the height of the fracture. Existing analyses [1-5] were based on the stress intensity factor solution for a crack in an isotropic and homogeneous elastic solid. The interfacial behavior was obtained as the limiting case when the distance between the crack tip and the material interface becomes vanishingly small. Since the stress intensity factor represents the amplitude of the inverse square root singularity in an isotropic and homogeneous material, and since the order of stress singularity for an interface crack is different from the inverse square root type, this limit in reality does not exist. This led to many misinterpretations of the results and generated confusion in the technical community. In an effort to clarify this situation, the present investigation addresses the containment problem based on the interfacial crack solution with the correct stress singularities. Since the physical meaning of the stress intensity factor for this case is unclear, a fracture criterion based on the maximum stress theory [7] was adopted for crack growth. Under these considerations, the physical process of hydraulic fracturing containment and the relative merit of the interfacial material properties and in situ stresses on containment will be discussed.

Although analytical models [8-10] exist which are capable of analyzing the interface crack geometry, these are not suitable for the complex geometries and loading conditions encountered in practice. Thus, another purpose of this investigation is to develop an efficient numerical tool such that parametric studies of the interactions between loading, material properties, and crack geometry involved in the massive hydraulic fracturing process can be carried out. The major difficulty associated with developing such a model lies in the treatment of the stress singularities for an interface crack. When the crack is embedded in a single material, stress intensifies at the crack tip as the order of the inverse square root of r, where r is the radial distance measured from the crack tip. When the crack tip approaches the material interface, the stress singularity is no longer of the inverse square root type, but changes to some value depending on the material properties associated with the interface. The enriched element method developed by Benzley [11] has been used by Chen [12] successfully to treat the problem of a crack lying along a material interface. Thus, for the current geometry in which a crack is terminated at a material interface, the enriched element method will again be adopted. The same geometry has also been treated by Lin and Mar [13], Cook and Tracey [14], and Atkinson and Javaherian [15] using various finite element techniques. Lin and Mar used the hybrid element method which includes a singular element at the crack tip. This element retains 16 terms of the infinite series expansion and, therefore, is rather complicated analytically. Cook and Tracey applied a three-node triangle element with embedded singularity. No stress intensity factor results were given in [14]. Atkinson and Javaherian modeled the interface problem as the limiting case when the crack tip extends very close to the interface from a single material. Thus, very small elements are required in the finite element idealization. The enriched element method for the present investigation models the crack tip singularity exactly and also calculates the stress intensity factors directly. It uses 12-node isoparametric elements which allows cubic displacement variations. As such, rather large elements can be used without affecting the calculation accuracy. In order to extend the enriched element method to interfacial cracking, crack tip displacement and stress expressions are required. These have been derived from the potential functions given by Cherepanov [16]. These expressions have been implemented on the finite element code PAPST [17]. PAPST is a special purpose fracture mechanics code which handles axisymmetric and planar geometries with a 12-node quadrilateral isoparametric element. Only the planar option has been modified in the present work. Good agreements between the present results and those available in the literature have been found. This indicates that the present method is an effective and efficient technique for studying interfacial cracking problems.

In the following sections, the development and validation of the finite element model will be presented. The model will then be used, in conjunction with the maximum stress criterion, to study the containment problem. It is found that the material properties influence crack growth only prior to the arrival of the crack tip at the interface. This effect is expressed in terms of the raising (if the material surrounding the cracked layer is stiffer) and lowering (for the opposite material combination) of the stress intensity factors such that the borehole pressure must be adjusted accordingly to insure crack growth. Once the crack tip reaches the interface, only the in situ stresses and interface slippage can retard the crack growth. Although the present model is not capable of assessing the interface slippage quantitatively, some qualitative observations are given.

CRACK TIP STRESS AND DISPLACEMENT EXPRESSIONS

As has been mentioned previously, the stress singularity for an interface crack is different from that for a crack embedded in a single material. This characteristic has been studied by many authors [18,8-10] and is well understood. For enriched element modeling purposes, explicit crack tip displacement and stress expressions are needed. However, aside from the work of Cherepanov [13], which gives partial stress fields, complete descriptions for these field variables are not available in the literature. In this section, the asymptotic crack tip displacement and stress expressions are derived. The derivation follows the potential function formulation given by Cherepanov [16].

Consider the two elastic half-planes of different materials in Figure 3, which are bonded together along the line $x = 0$. A semi-infinite crack exists in the left half-plane whose tip terminates at the material interface. The material occupying the left half-plane has Young's modulus E_1 and Poisson's ratio ν_1, while those for the right half-plane are E_2 and ν_2, respectively. The plane elasticity problem can be formulated in terms of four complex functions $\Phi_j(z)$ and $\Psi_j(z)$ of the complex variable $z = x + iy$, where $j = 1,2$ refers to quantities associated with the material 1 and 2, respectively. The stress and displacement fields are related to the complex functions in the Kolosov-Muskhelishvili form as

$$(\sigma_x)_j + (\sigma_y)_j = 2\,[\Phi_j(z) + \overline{\Phi_j(z)}] \tag{1}$$

$$(\sigma_y)_j - (\sigma_x)_j + 2\,i\,(\tau_{xy})_j = 2\,[\bar{z}\,\Phi_j(z) + \Psi_j(z)] \tag{2}$$

$$2\,G_j\,(u_j + i\,v_j) = \kappa_j \phi_j(z) - z\,\overline{\Phi_j(z)} - \overline{\psi_j(z)} \tag{3}$$

$$\Psi_j(z) = \psi_j(z) = d\,\psi_j/dz,\quad \Phi_j(z) = d\,\phi_j(z)/dz \tag{4}$$

where u_j and v_j are components of displacements in the x and y directions, respectively, while the stresses conventions are given in Figure 1. In the above equations, G_j is the shear modulus of elasticity and κ_j is a material constant where $\kappa_j = 3 - 4\nu_j$ for plane strain and $\kappa_j = (3 - \nu_j)/(1 + \nu_j)$ for plane stress. Also the bar superscripts operating on a function, say $\bar{\Phi}_j$, denotes the complex conjugate of Φ_j.

Following Cherepanov [16], the complex potentials can be represented by the following expressions:

$$\Phi_j(z) = A_j\, z^{\lambda} \tag{5}$$

$$\Psi_j(z) = B_j\, z^{\lambda} \tag{6}$$

in which λ, A_j and B_j are complex constants to be determined from the symmetry, continuity and boundary conditions of the problem. By enforcing these conditions, the constants A_j and B_j are given as [16]

$$A_1 = A_2\,\{\cos\frac{\pi\lambda}{2}\,[k_2 - k_1\,(\lambda + 2 - k_3)] + i\,\sin\frac{\pi\lambda}{2}\,[k_2 + k_1(\lambda - k_3)]\} \tag{7}$$

$$B_1 = A_2\,\{\cos\frac{\pi\lambda}{2}\,[(\lambda + 2)<1 - k_2 + k_1(\lambda + 2)> - <1 + k_1(\lambda + 2)>k_3]$$

$$-i\,\sin\frac{\pi\lambda}{2}\,[\lambda\,(\lambda\,k_1 + k_2 - 1) + (1 - \lambda\,k_1)k_3]\} \tag{8}$$

$$A_2 = K_I\ (\lambda+1)/\ (2\pi)^{1/2} \tag{9}$$

$$B_2 = A_2\ k_3 \tag{10}$$

in which K_I is the stress intensity factor, and the contractions

$$k_1 = (G-1)/(\kappa_1+1), \qquad k_2 = G(\kappa_2+1)/(\kappa_1+1), \qquad G = G_1/G_2 \tag{11}$$

$$k_3 = [k_1(3\lambda+2)-k_2(1+2\lambda)+\lambda+1]/\ (k_1+1) \tag{12}$$

have been made. The eigenvalue λ is to be determined from the following characteristic equation:

$$\cos\pi\lambda = a+b(\lambda+1)^2, \qquad \lambda\min\ \varepsilon\ [-1,0] \tag{13}$$

$$a = (2k_1^2-2k_1k_2+2k_1-k_2+1)/[2(k_1+1)(k_2-k_1)] \tag{14}$$

$$b = 2k_1/(k_1+1) \tag{15}$$

Note that the lowest root of λ from this equation is real and it represents the stress singularity at the crack tip. Based on the above, the asymptotic stress and displacement fields are found as

$$(\sigma_x)_1 = A_2r^\lambda\ \{\cos\frac{\pi\lambda}{2}\ [(2c_1+c_3)\cos\lambda\Theta + \lambda c_1\cos(\lambda-2)\Theta\ ] \\ -\sin\frac{\pi\lambda}{2}\ [(2c_2-c_4)\sin\lambda\Theta + \lambda_2\ \sin(\lambda-2)\Theta\ ]\} \tag{16}$$

$$(\sigma_y)_1 = A_2r^\lambda\ \{\cos\frac{\pi\lambda}{2}\ [(2c_1-c_3)\cos\lambda\Theta - \lambda c_1\cos(\lambda-2)\Theta\ ] \\ -\sin\frac{\pi\lambda}{2}\ [(2c_2+c_4)\sin\lambda\Theta - \lambda c_2\sin(\lambda-2)\Theta\ ]\} \tag{17}$$

$$(\tau_{xy})_1 = A_2r^\lambda\ \{-\cos\frac{\pi\lambda}{2}\ [\lambda\ c_1\sin(\lambda-2)\ \Theta+c_3\sin\lambda\Theta\ ] \\ -\sin\frac{\pi\lambda}{2}\ [\lambda\ c_2\cos(\lambda-2)\ \Theta-c_4\cos\lambda\Theta\ ]\} \tag{18}$$

$$u_1 = -[A_2/(\lambda+1)][r^{\lambda+1}/(2G_1)]\{\cos\frac{\pi\lambda}{2}\ [(\kappa_1\ c_1+c_3)\sin(\lambda+1)\Theta \\ +(\lambda+1)c_1\sin(\lambda-1)\Theta\ ]+\sin\frac{\pi\lambda}{2}\ [(\kappa_1\ c_2-c_4)\cos(\lambda+1)\Theta \\ +(\lambda+1)c_2\cos(\lambda-1)\Theta\ ]\} \tag{19}$$

$$v_1 = [A_2/(\lambda+1)][r^{\lambda+1}/(2G_1)]\{\cos\frac{\pi\lambda}{2}\ [(\kappa_1\ c_1-c_3)\cos(\lambda+1)\Theta \\ -(\lambda+1)c_1\cos(\lambda-1)\Theta\ ]-\sin\frac{\pi\lambda}{2}\ [(\kappa_1\ c_2+c_4)\sin(\lambda+1)\Theta \\ -(\lambda+1)c_2\sin(\lambda-1)\Theta\ ]\} \tag{20}$$

where $\Theta = \theta - \pi/2$, and

$$(\sigma_x)_2 = A_2 r^{\lambda} \; [(2-k_3)\cos\lambda\theta \;\; - \lambda\cos(\lambda \; -2)\theta \;] \tag{21}$$

$$(\sigma_y)_2 = A_2 r^{\lambda} \; [(2+k_3)\cos\lambda\theta \;\; + \lambda\cos(\lambda \; -2)\theta \;] \tag{22}$$

$$(\tau_{xy})_2 = A_2 r^{\lambda} \; [k_3 \sin\lambda\theta \;\; + \lambda\sin(\lambda \; -2)\theta \;] \tag{23}$$

$$u_2 = [A_2/(\lambda \; +1)][r^{\lambda+1}/(2G_2)][(\kappa_2 \; -k_3)\cos(\lambda \; +1)\theta$$
$$-(\lambda \; +1)\cos(\lambda \; -1)\theta \;] \tag{24}$$

$$v_2 = [A_2/(\lambda \; +1)][r^{\lambda+1}/(2G_2)][(\kappa_2 \; +k_3)\sin(\lambda \; +1)\theta$$
$$+(\lambda \; +1)\sin(\lambda \; -1)\theta \;] \tag{25}$$

In the above equations, r and θ represent the polar coordinates originated at the crack tip, Figure 3, and the following simplifications

$$c_1 = k_2 - k_1(\lambda \; +2) + k_1 k_3 \tag{26}$$

$$c_2 = k_2 + k_1\lambda \; - k_1 k_3 \tag{27}$$

$$c_3 = (\lambda \; +2)[1-k_2+k_1(\lambda \; +2)] - (1+k_1\lambda \; +2k_1)k_3 \tag{28}$$

$$c_4 = \lambda \; (\lambda \; k_1 + k_2 - 1) + (1-\lambda \; k_1)k_3 \tag{29}$$

have been made. Equations 16–25 completely characterize the near field stresses and displacements to be used in the enriched element formulations. From Equation 22, it is seen that along the line $\theta = 0$, the normal stress reduces to

$$(\sigma_y)_2 \; (r,0) = [\; K_I(\lambda \; +1)/(2\pi \;)^{1/2} \;] \; r^{\lambda} \; (\; 2+k_3+ \; \lambda) \tag{30}$$

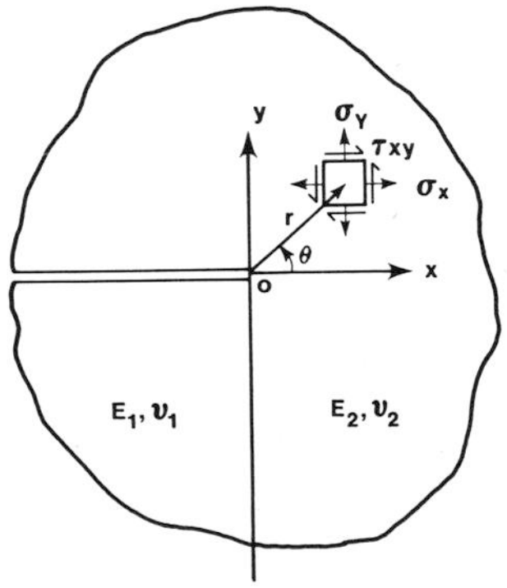

Figure 3. Interface Crack Geometry

Equation 30 provides the relationship between the stress intensity factor K_I defined in this manner and that in References 8–10.

Code implementation has been accomplished on the program PAPST [24]. The procedure follows that given in [12] and the details will not be repeated here. Validation of the calculation results against analytical and other finite element results has been given in [24], and the enriched element method has proven to be an accurate and efficient tool for the analysis of interface crack problems.

CONTAINMENT PROBLEM

Consider the simplified plain strain geometry given in Figure 2. A layer of width 2h, Young's modulus E_1 and Poisson's ratio ν_1 is sandwiched between two half-planes of materials properties E_2 and ν_2. The layer contains a crack through the width whose surfaces are being pressurized. The boundaries of the layered medium are subjected to the tractions σ_v, σ_1, σ_2, as shown in Figure 2. The problem is to determine the condition of crack extension under various material and loading combinations by combining the displacement and stress solutions with an appropriate fracture criterion.

Before the discussion on the interface crack problem, a few remarks on the behavior prior to the arrival of the crack tip at the interface are in order. In this case, the crack is completely embedded in an isotropic and homogeneous material; and the stress intensity factor is a valid parameter governing crack growth, provided the crack tip is not too close to the interface. The stress intensity factor solutions for this geometry have been obtained by [19,20] and summarized in [21]. The characteristic of the stress intensity factor as a function of the distance of the crack tip from the interface is depicted schematically in Figure 4. It is observed that the stress intensity factor increases with decreasing distance between the crack tip and the interface when $E_1/E_2 > 1$. The opposite is true when $E_1/E_2 < 1$. This implies that if everything else being equal, more (less) pressure is required to extend the crack when the surrounding medium is stiffer (softer) than the material in the cracked layer. Bear in mind, however, that this stress intensity factor represents the amplitude of the inverse square root stress singularity. As the crack tip approaches the interface, this stress singularity is changed from the inverse square root type to some other value determined from the characteristic equation (13). This indicates that the stress intensity factor as defined here is not valid at distances very close to the interface. Indeed, it is seen from Figure 4 that the stress intensity factor increases without bound at the interface for $E_1 > E_2$ and decreases to zero for $E_1 < E_2$, which are physically unrealistic. This behavior has been interpreted as the capability of the material property difference to contain crack extension at the interface, i.e.,

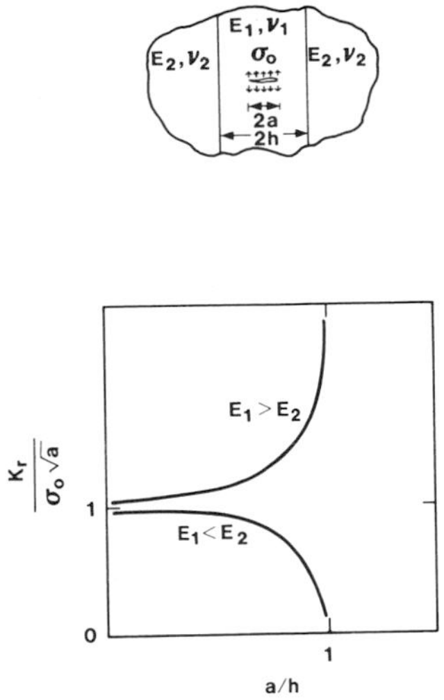

Figure 4. Embedded Crack in a Layer of Material

if the surrounding material is stiffer, the crack will be arrested at the interface while no containment will result if the surrounding material is softer. Ample experimental evidence exists to show that this is not the case. Limiting the range of validity of the curves in Figure 4 to some distance from the interface, our interpretation would be that the material property difference determines the level of pressure required to maintain crack extension toward the interface. In other words, higher pressure is required to extend the crack toward a stiffer material than a softer material. If sufficient pressure can be maintained in the borehole, the crack tip will eventually arrive at the interface regardless of the material property difference associated with the interface.

FINITE ELEMENT SOLUTION

Now let us return to the interface crack problem whose geometry is shown in Figure 2. Because of symmetry, only a quarter of the layered medium needs to be considered in the finite element solution. Figure 5 depicts the finite element mesh for this problem. Note that the geometry has been rotated 90 degrees from that in Figure 2. This mesh includes 56 elements. By taking $\nu_1 = \nu_2 = 0.2$, four E_1/E_2 ratios of 10, 2, 0.5, and 0.1 are included in this study. The crack surface is pressurized by σ_o and results for several combinations of the in situ stresses have been obtained. These results are shown in Figures 6-8 in terms of the normal stress ahead of the crack tip, $(\sigma_y)_2\ (r,0)$. Figures 6-8

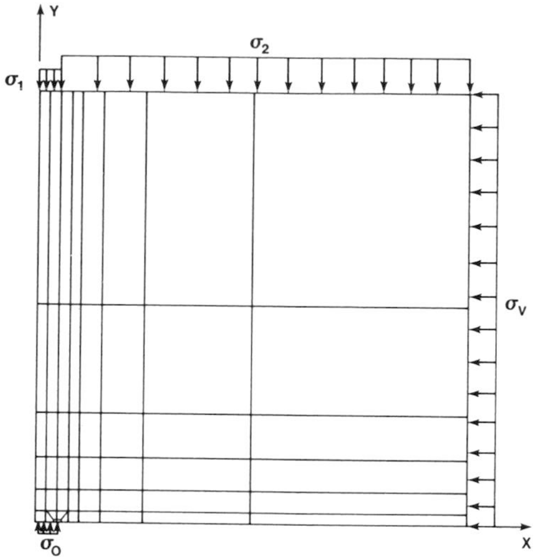

Figure 5. Finite Element Mesh for Containment Problem

show the normal stress as a function of the distance from the interface and the Young's modulus ratio for the in situ stress combinations $\sigma_v = 0$, and $(\sigma_1/\sigma_o, \sigma_2/\sigma_o)$ of (0,0), (0, 0.5), and (0,1.0), respectively. These results are important for the application of the fracture criterion to be introduced in the next section. A point associated with interface slippage concerns the relationship between the in situ stresses. From [10], it is shown that these stresses are not independent of each other in order for the stresses and displacements to be continuous across the interface. The relationship is given as [10]

$$G_2 (1-\nu_1) \sigma_1 = (G_2 \nu_1 - G_1 \nu_2) \sigma_v + G_1 (1-\nu_2) \sigma_2 \tag{31}$$

Clearly, this equation is in general not satisfied by the in situ stress combinations in the field. This implies that the bonding is not perfect and slippage of the interface occurs as a rule rather than the exception. If the site of slippage coincides with the expected path of crack extension, the crack can conceivably be arrested at the interface. Unfortunately, the current model is not capable of quantifying this observation.

MAXIMUM STRESS FRACTURE CRITERION

Since the stress singularity changes with the material property combinations, the stress intensity factor for an interface crack as defined by

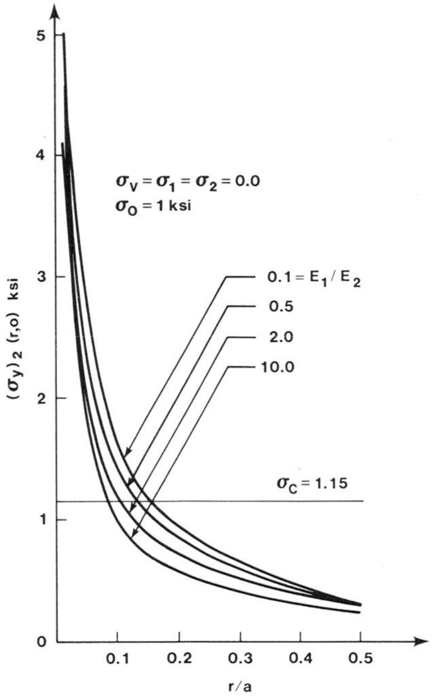

Figure 6. Stress vs. Position Plots for $\sigma_2 = 0.0$

Equation 9 possesses different physical dimensions from case to case. Thus, it is not suitable to serve as a fracture parameter. In order to determine the in situ stress effect on fracture containment, a fracture criterion based on the maximum stress ahead of the crack tip and along the expected path of crack extension is adopted. This criterion was proposed by Erdogan [22] for nonhomogeneous materials and has been shown in [23] to be reasonable over a range of fiber reinforced composites.

In terms of the coordinate system in Figure 5, this is expressed as

$$(\sigma_y)_2\ (\delta,\ 0) = \sigma_c \tag{32}$$

where σ_c is the tensile strength of material 2 and δ is a length parameter. Both σ_c and δ are considered as properties of the material. The parameter δ depends on the microstructure of the crack tip zone which is very complex. One way to determine it is to examine the crack tip stresses in an isotropic and homogeneous material. The crack tip stress can be written as

$$\sigma_y (x, 0) = K_I / (2 \pi x)^{1/2} \qquad (33)$$

According to the stress intensity factor criterion, crack extension initiates when K_I reaches K_{IC}. Thus, Equation 33 becomes

$$\sigma_y (x, 0) = K_{IC} / (2 \pi x)^{1/2} \qquad (34)$$

However, from Equation 32, the crack extension condition can alternatively be expressed as

$$\sigma_c = K_{IC} / (2\pi\delta)^{1/2} \qquad (35)$$

Thus, δ can be found from Equation 35 if the fracture toughness and the tensile strength of the material are known. For a typical shale material with tensile strength of 7.9 MPa (1.15 ksi) and fracture toughness of 1 $MN/m^{3/2}$ (0.91 ksi $in^{1/2}$), δ/h is determined as 0.1 where h is the half width of the cracked layer, or the crack length. This value will be used for the present parametric studies.

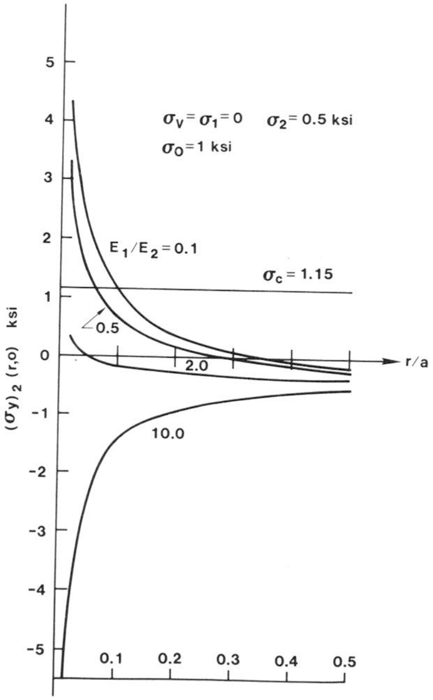

Figure 7 Stress vs Position Plots for $\sigma_2 = 0.5$

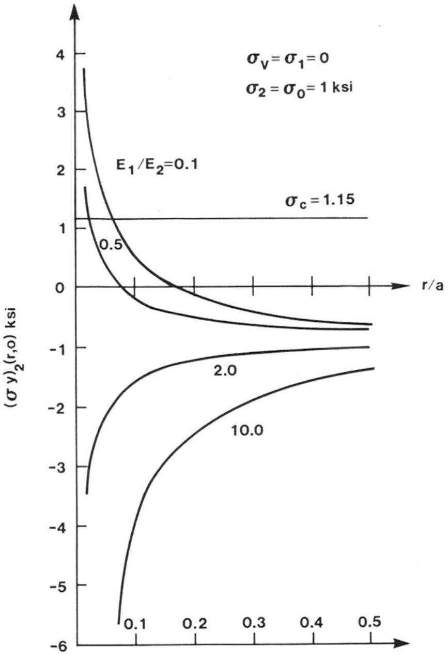

Figure 8 Stress vs. Position Plots for $\sigma_2 = 1.0$

Combining the results in Figures 6–8 and the Maximum Stress Criterion, several observations can be made. Without the application of the in situ stresses, crack extension through the interface will occur for E_1/E_2 ratios of 0.1, 0.5, and 2.0 but not for $E_1/E_2 = 10$. This is surprising since the only containment case corresponds to one in which the crack is extending into a softer material. Also, in general, higher stress is produced in the combination of a crack growing into a stiffer material. An in situ stress difference, $(\sigma_2 - \sigma_1)$, of half of the pressure, σ_o, applied to the crack surface is enough to suppress crack extension through the interface for all material combinations. Moreover, if the in situ stress difference is increased to σ_o, the tensile stress ahead of the crack tip is further reduced. Based on these results, it can be concluded that the in situ stress is the dominant parameter in containing the fracture growth through the material interface when the crack tip has reached the material interface. It must be emphasized that these results have been reached by considering only one crack extension mode. The case where the crack grows along the interface, or interface slippage, has not been included.

CONCLUSIONS

The enriched element method has been applied to study the stress analysis involving a bimaterial crack oriented in a direction perpendicular to the material interface. Explicit crack tip stress and displacement expressions were derived to facilitate this application. These expressions were coded into the finite element program PAPST to perform the analysis. For the range of material properties and loading and geometry combinations included in this study, this technique appears to offer an accurate numerical tool for the analysis of bimaterial interface crack problems. This technique was applied to determine the relative merit of in situ stress and material property differences in the containment of hydraulic fractures. The current approach differs from previous efforts in that the inability of the classical stress intensity factor criterion to deal with the interfacial cracking geometry was recognized from the outset. Therefore, a crack growth criterion based on the maximum tensile stress at a fixed distance ahead of the crack tip was defined. The finite element solution is particularly suited for this fracture criterion since the correct stress singularity has been included in the interface crack tip stress field. The results indicate that the in situ stresses appear to be the dominating parameter in prohibiting the crack to extend through the material interface once the crack tip has arrived at the interface. The material property differences play a role in determining the pressure level required for the crack tip to reach the interface. This analytical result confirms the experimental observations reported by Warpinski, Schmidt, and Northrop [6].

REFERENCES

1. E. R. Simonson, A. S. Abou-sayed, and R. J. Clifton, Containment of Massive Hydraulic Fracture, Society of Petroleum Engineers Journal, pp. 27–32, February, 1978.
2. A. A. Daneshy, Hydraulic Fracture Propagation in Layered Formation, Society of Petroleum Engineers Journal, pp. 33–41, February, 1978.
3. M. E. Hanson and R. J. Shaffer, Some Results From Continuum Mechanics Analyses of the Hydraulic Fracturing Process, Society of Petroleum Engineers Journal, pp. 86–94, April, 1980.
4. M. P. Cleary, Primary Factors Governing Hydraulic Fractures in Heterogeneous Stratified Porous Formations, ASME Paper No. 78-pet-47, presented at the Energy Technology Conference and Exibition, Houston, Texas, November 5–9, 1978.
5. J. C. Roegiers and T. D. Wiles, Hydraulic Fracturing of Rock Masses, Earth Physics Branch Report, Geothermal Service of Canada, Ottawa, Canada, 1981.
6. N. R. Warpinski, R. A. Schmidt, and D. A. Northrop, In Situ Stresses: The Predominant Influence on Hydraulic Fracture Containment, Paper SPE 8932, Presented at the 1980 SPE/DOE Symposium on Unconventional Gas Recovery, Pittsburgh, Pennsylvania, May 18–21, 1980.
7. F. Erdogan, Fracture of Nonhomogeneous Solids, THE MECHANICS OF FRACTURE, ASME publication AMD – Vol. 19, pp. 155–170, 1976.

8. T. S. Cook and F. Erdogan, Stresses in Bonded Materials With a Crack Perpendicular to the Interface, International Journal of Engineering Science, Vol. 10, pp. 677-697, 1972.
9. G. D. Gupta, A Layered Composite With a Broken Laminate, International Journal of Solids and Structures, Vol. 9, pp. 1141-1154, 1973.
10. N. E. Ashbaugh, Stresses in Laminated Composites Containing a Broken Layer, ASME paper 72-WA/APM-14, Presented at the 1972 ASME Winter Annual Meeting, New York, New York, November 26-30, 1972.
11. S. E. Benzley, Representations of Singularities With Isoparametric Finite Elements, International Journal for Numerical Methods in Engineering, Vol. 8, pp. 537-545, 1974.
12. E. P. Chen, Finite Element Analysis of a Bimaterial Interface Crack, Theoretical and Applied Fracture Mechanics, Vol. 3, pp. 257-262, 1985.
13. K. Y. Lin and J. W. Mar, Finite Element Analysis of Stress Intensity Factors for Cracks at a Bimaterial Interface, International Journal of Fracture, Vol. 12, pp. 521-531, 1976.
14. T. S. Cook and D. M. Tracey, Stress Distribution in a Cracked Bimaterial Plate, FRACTURE 1977, Proceedings of the Fourth Congress on Fracture, Vol. III, pp. 1055-1058, University of Waterloo Press, Waterloo, Canada.
15. C. Atkinson and H. Javaherian, Evaluation of Stress Intensity Factors Associated With Bimaterial Cracks, Journal of the Institute of Mathematics Applications, Vol. 26, pp. 235-258, 1980.
16. G. P. Cherepanov, MECHANICS OF BRITTLE FRACTURE, McGraw-Hill Inc., 1979.
17. L. N. Gifford and P. D. Hilton, Preliminary Documentation of PAPST - Nonlinear Fracture and Stress Analysis by Finite Elements, David W. Taylor Naval Ship Research and Development Center, Bethesda, Maryland, 1981.
18. A. R. Zak and M. L. Williams, Crack Point Stress Singularities at a Bimaterial Interface, Journal of Applied Mechanics, Vol. 30, pp. 142-143, 1963.
19. P. D. Hilton and G. C. Sih, Laminated Composite With A Crack Normal to the Interfaces, International Journal of Solids and Structures, Vol. 7, pp. 913-930, 1971.
20. F. Erdogan and G. D. Gupta, Stress Analysis of Multilayered Composites With a Flaw, International Journal of Solids and Structures, Vol. 7, pp. 39-61, 1971.
21. G. C. Sih and E. P. Chen, CRACKS IN COMPOSITE MATERIALS, Noordhoff International Publishing, The Netherlands, 1981.
22. E. Edogan, Fracture of Nonhomogeneous Solids, The Mechanics of Fracture, AMD-Vol. 19, pp. 155-170, The American Society of Mechanical Engineers, 1976.
23. G. Caprino, J. C. Halpin and L. Nicolais, Fracture Mechanics in Composite Materials, Composites, pp. 223-227, October, 1979.
24. E. P. Chen, Finite Element Analysis of Bimaterial Interface Cracks, Advances in Aerospace Science and Engineering, U. Yuceogln and R. Hesser, editors, ASME Publication AD-08, pp. 105-111, 1984.

AUTHOR INDEX

SUBJECT INDEX

This index is compiled from key words submitted by the authors and supplemented by the Editors; the numbers refer to the first page of the paper concerned.